THERMODYNAMICS OF SPONTANEOUS AND NON-SPONTANEOUS PROCESSES

THERMODYNAMICS OF SPONTANEOUS AND NON-SPONTANEOUS PROCESSES

I. M. KOLESNIKOV, S. I. KOLESNIKOV,
V. A. VINOKUROV AND
G. E. ZAIKOV

Nova Science Publishers, Inc.
Huntington, New York

Senior Editors: Susan Boriotti and Donna Dennis
Office Manager: Annctte Hellinger
Graphics: Wanda Serrano
Information Editor: Tatiana Shohov
Book Production: Cathy DeGregory, Kay Seymour, Jennifer Vogt and Lynette Van Helden
Circulation: Ave Maria Gonzalez, Ron Hedges, Andre Tillman

Library of Congress Cataloging-in-Publication Data
Available upon request.

Includes bibliographical references and index.
ISBN 1-56072-904-X

227 Main Street, Suite 100
Huntington, New York 11743
Tele. 631-424-6682 Fax 631-424-4666
e-mail: Novascience@earthlink.net
Web Site: http://www.nexusworld.com/nova

Printed in the United States of America

CONTENTS

Abstract ix

Editor's Preface ix

Authors' Preface x

About the Author xiii

CHAPTER 1 MAIN TERMS IN THERMODYNAMICS 1

1.1. Introduction 1

1.2. Terms of Systems and Processes 3

1.3. Thermodynamic Parameters and Functions 7

1.4. Classification of the Matter Objects and Forms of Motion 10

1.5. Time Parameter in Thermodynamics 12

1.6. The Law of Conservation in Thermodynamics of Irreversible Processes 15

1.7. Vector-Scalar Change of Energy 17

CHAPTER 2 SPONTANEOUS AND NON-SPONTANEOUS PROCESSES 19

2.1. Idea about Spontaneous and Non-Spontaneous Processes 19

2.2. Self-Organization as a Consequence of Influence of Non-Spontaneous Processes on Structure, Composition and State of the System 20

CHAPTER 3 THE FIRST LAW OF THERMODYNAMICS FOR OPEN SYSTEMS 27

3.1. Formulation and Analytical Expression 27

3.2. Work and Heat in Spontaneous and Non-Spontaneous Processes 28

3.3. Internal Energy 30

3.4. General Formulation of the First Law of Thermodynamics. Heat Capacity 31

3.5. Enthalpy in Open Systems 34

3.6. Change of Enthalpy and Thermochemistry as a Section of Thermodynamics. Designations in Thermochemistry 38

3.7. The Law by G.I. Gess 41

3.8. Temperature Dependency of the Heat Effect 44

CHAPTER 4 THE SECOND LAW OF THERMODYNAMICS AND MAIN MATHEMATICAL EQUATIONS 49

4.1. Insufficiency of the First Law of Thermodynamics to Classify Processes 49

4.2. Spontaneous and Non-Spontaneous Processes 50

4.3. Illustration of a Pair of Conjugated Processes in the Frames of a Single System 53

4.4. Reversible and Irreversible Processes 54

4.5. Formulation of the Second Law of Thermodynamics 57

4.6. Analytical Determination of Entropy 62

4.7. Temperature Scale as the Level of Entropy Reading 73

4.8. Generalized or the Main Equation of Thermodynamics for Equilibrium and Non-Equilibrium Processes. 78

CHAPTER 5 THERMODYNAMICS OF SPONTANEOUS AND NON-SPONTANEOUS PROCESSES 83

5.1. Necessity of Development of a New Direction in Thermodynamics 83

5.2. Spontaneous and Non-Spontaneous Processes on Different Levels of State of the Matter 84

5.3. Phase-Open Systems 85

5.4. Entropy of Thermodynamically Irreversible Spontaneous and Non-Spontaneous Processes 87

5.5. Thermodynamic Proof of the Ratio of Inequalities for Spontaneous and Non-Spontaneous Processes 92

5.6. Changing of Entropy in a Phase-Open System Interacting with the Surrounding 95

CHAPTER 6 CORRELATION OF PROCESSES FOR INTERACTING PHASE-OPEN SYSTEMS AND THE SURROUNDING 99

6.1. Direction of the Process Proceeding 99

6.2. The Example of Heat Processes with Diminution and Growth of Entropy 100

6.3. Interaction of System, Including Subsystems, with the Surrounding 103

CHAPTER 7 KINETICS OF ENTROPY VARIATION 105

7.1. Main Concepts 105

7.2. Entropy of Substance Flow in the Open System 106

7.3. General Equation of the Heat Flow 112

7.4. Entropy Flow 113

CHAPTER 8 THE HELMHOLTZ ENERGY FOR SPONTANEOUS AND NON-SPONTANEOUS PROCESSES 117

8.1. Analytical Expression of the Helmholtz Energy 117

8.2. Relation Between the Helmholtz Energy and Temperature and Volume 119

8.3. Relation Between the Helmholtz Energy, Heat and Work 120

8.4. Criteria Determining the Type of the Process Proceeding 121

8.5. The Helmholtz Energy Including Heat of Irreversible – Spontaneous and Non-Spontaneous – Processes 123

8.6. Connection of the Helmholtz Energy with Heat and Work 124

8.7. Analytical Expression of the Helmholtz Energy for Open Systems 125

CHAPTER 9 THE GIBBS ENERGY IN THERMODYNAMICALLY IRREVERSIBLE PROCESSES 129

9.1. Analytical Expression for the Gibbs Energy for Conditions of P and T Constancy in Chemical and Phase Homogeneous System 129

9.2. Connection of the Gibbs Energy with Pressure and Temperature 132

9.3. Relation Between the Gibbs Energy, Heat and Work 134

9.4. Changing of the Gibbs Energy in Isolated Systems 135

9.5. Criteria of Types of Processes Proceeding in Open Systems 136

9.6. The Gibbs Energy and its Connection with Heats of the System and the Surrounding 138

9.7. Connection of the Gibbs Energy to Heat and Work 139

9.8. Changing of the Gibbs Energy for Open Systems with Mass Exchange and Chemical Reactions. Analytical Expression of the Gibbs Energy with External and Internal Mass Exchange 139

9.9. Analytical Expression of the Gibbs Energy for Open Systems with Mass Exchange Between Phases and the Surrounding 140

9.10. The Gibbs Energy for Open Systems with Mass Exchange and Chemical Reactions 141

9.11. Conclusion to the Theoretical Part 142

CHAPTER 10 PRACTICAL EXAMPLES OF INFLUENCE OF RELATION OF SPONTANEOUS AND NON-SPONTANEOUS PROCESSES ON TECHNOLOGICAL AND NATURAL PHENOMENA 145

10.1. Dispersion of the Continuum 145

10.2. Thermal Machines 147

10.3. Mechanical Processes 148

10.4. Catalytic Processes 148

CHAPTER 11 EQUATIONS OF EQUILIBRIUM THERMODYNAMICS AND THE METHOD OF DETERMINATION OF THE PROCESS TYPE BASING ON THERMODYNAMICS OF SPONTANEOUS AND NON-SPONTANEOUS PROCESSES 149

11.1. Equations of the First Law of Thermodynamics and their Connection with the Process Type 149

11.2. The Equation of the First Law of Thermodynamics Determining Heat Delivery to the Working Substance 150

11.3. Equation of the First Law of Thermodynamics for Spontaneous Process 151

11.4. Changing of the Helmholtz Energy for Spontaneous and Non-Spontaneous Processes 152

References 155

INDEX 157

ABSTRACT

The book is devoted to development of a new direction of irreversible processes in thermodynamics — thermodynamics of spontaneous and non–spontaneous processes. Formulae of the first and the second laws of thermodynamics are discussed in a new context. The theorem about interaction of spontaneous and non–spontaneous processes are formulated and proved, mathematical expressions for thermodynamic functions is displayed, criteria of determination of the process proceeding direction and equilibrium conditions are formulated. Examples of interactions between spontaneous and non–spontaneous processes are discussed.

The book will be suitable for students, post–graduates, scientists and professors of physicochemical specialization.

EDITOR'S PREFACE

"Russian people…
will bear everything,
and broad and clear
pathway will make
by their breasts…
unfortunately, neither
you, nor me will
live up to these days"
Nikolai A. Nekrasov
Russian poet,
middle XIX

The words, put to the epigraph by me, clearly display the difference between kinetics and thermodynamics. Kinetics represents the science about the rate of chemical reactions, about dynamics of processes. It answers the question "When?" or "How fast?" At the same time, thermodynamics is the science about equilibrium in chemical processes. It does not say how fast this equilibrium will be reached, but just states this equilibrium.

The poem by Nekrasov puts forward the indisputable fact that Russian people will live very well (this is thermodynamics). The same was told in propagandistic speeches in the former USSR and is told now in pseudo-democratic times in this country. However, the question is risen: " When will it be?" This is the question of kinetics. One and half centuries ago Nekrasov stated that he will not live up to this "happy future", and he was right. I am to state that the present generation will not live up to these "good days", which must appear some day in this country. However, even under conditions far from the ideal Russian science exists and develops.

The present monograph on thermodynamics is the example of this. Her point of view of Professor Ivan M. Kolesnikov and his colleagues on some questions of thermodynamics is discussed. For more than 40 years, I.M. Kolesnikov himself is working as the professor of physical chemistry in Moscow Gubkin's Academy of Oil and Gas. He is one of the leading scientists of our country in the field of thermodynamics, kinetics and catalysis in oil chemical processes. This book will be the first monograph, published in the USA. If it has a success, we will continue publication of books by school of Professor I.M. Kolesnokov for English speaking readers.

We will be greatly thankful for any constructive notes and suggestions of readers to the authors and the editor of this monograph.

Prof. Guennadi E. Zaikov – Editor
N.M. Emanuel Institute of Biochemical Physics, Russian Academy of Sciences
4, Kosygin str., Moscow 117334, Russia
Fax: (7-095) – 137-4101
E–mail: chembio@glasnet.ru
Tel: (7-095) – 939-7191

AUTHORS' PREFACE

At the present time two methods are developed to study proceeding of natural and technological thermodynamic processes: the hypothetical–logic method of equilibrium thermodynamics and the method of irreversible process thermodynamics applied to really proceeding processes. Methods of the equilibrium thermodynamics in studying thermodynamic processes play the role of border theorems and enable parameters of initial and end states of systems to be calculated. These states are reached in border cases, that is, during the system transition from one level to another taking into account no intermediate stages and the mechanism of their proceeding.

Thermodynamics of equilibrium systems allows state of the system to be determined precisely for equilibrium conditions of their existence.

Thermodynamics of irreversible processes logically and mathematically studies the behavior of physicochemical processes in open and closed systems (its methods are also spread upon isolated systems) during their spontaneous proceeding (cooling of heated up bodies, evaporation, diffusion, dissolving, pressure decreasing, and others). Consequently, it encloses just one of aspects of proceeding and changing of the system state in thermodynamically irreversible processes. The above–mentioned does not mean that the methods of thermodynamics of irreversible processes are incorrect, this note being probably related to the entropy growth principle in thermodynamically irreversible processes. The latter just means that up to now the main attention in thermodynamics of irreversible processes was adhered to spontaneous ones only, to say, the one–sided approach to the study of regularities in proceeding of real natural and technological processes was performed.

In this case, an idea of necessary broadening of the methods of irreversible processes' thermodynamics suggests itself, the list of methods including such as those considering participation of non–spontaneous processes in proceeding of the real ones under interaction between open and closed systems. This will enable a definite uncertainty to be excluded from our relationship with the environment, laws, principles and postulates of the classic thermodynamics and that of irreversible processes to be refined. The everyday observation (this observation is indifferent for the nature) of the environment allows a notice that it is the center point of the interaction of spontaneous and non–spontaneous processes: heating—cooling, pressure boosting in one part of the atmosphere—pressure decrease in another part of it (cyclone—anticyclone), evaporation—condensation, body falling in the Earth gravitation field—body lifting, and the infinite number of other pairs of interacting spontaneous and non–spontaneous processes. This approach will probably enable the necessity to adhere the preference to systems with the end state to be eliminated in the general scale of problems and the interaction of systems with the environment to be considered and, vice versa, in their real interaction taking into account extent of the systems and their lifetime (time itself being applied only as an increment of the parameter extent, paying no special attention to beginning and end for the general understanding of time).

It should be probably postulated that in the environment (it can be named absolutely different — the Universe, Para–universe, Hyper–universe, although such terms are not quite fortunate, because of their somewhat definition of finite size of defined objects and finiteness of their lifetime) spontaneous and non–spontaneous processes proceed simultaneously and continuously, interacting with each other and affecting each other (the environment and systems).

The number of such systems and pairs of processes in the environment is not definite, and for the number of such systems and pairs of processes proceeding in them, conjugated to the environment, it is senseless to use finite categories. Consequently, all this must be perceived in the way as it is presented by the environment and the nature of interacting processes, reflecting it — as a combination of spontaneous and non–spontaneous processes.

Development of methods of such approach to the study of events in the nature and technology is something new in thermodynamics. That is why the authors are unable to eliminate all mistakes in their presentation and deduction of formulations and laws in a new direction of development of thermodynamics. This will be offensive for the authors. However, on mistakes of scientists the mankind learns the nature and processes proceeding in it, and correction of mistakes is the effective instrument of development of the science about nature.

The present book is the first attempt to create a new method to study thermodynamics of irreversible processes taking into account pairs of spontaneous and non–spontaneous processes.

Senior scientific investigator S.I. Kolesnikov
Professor V.A. Vinokurov
Professor I.M. Kolesnikov

ABOUT THE AUTHORS

Kolesnikov Ivan Mikhailovich, Professor – was born in 1929, July 1 in Fashchevatoe village of Korochansky district, Belgorodsky region. He studied as a student of I.M. Gubkin Oil Institute of Moscow during 1949–1954. He was PhD graduated in 1959, and obtained the degree of Doctor of Sciences on the theme "Catalysis on minerals from the group of *ortho*-silicates and synthetic alumosilicates". Since 1970 I.M. Kolesnikov works as a professor in the department of physical and colloid chemistry of I.M. Gubkin Russian State University of Oil and Gas. He is the scientific leader of the branch laboratory of industrial kinetics and catalysis.

I.M. Kolesnokov is the real member of New York Academy of Sciences. His main branches of interest in science are: kinetics and catalysis, chemical thermodynamics of spontaneous and non-spontaneous processes, modeling of oil processing processes, quantum chemistry applied to catalysis.

I.M. Kolesnikov is the creator of the theory of catalysis by polyhedrons. He has formulated the generalized quantum-chemical principle, develops thermodynamics of spontaneous and non-spontaneous processes. He published over 620 scientific works, including 16 monographs, 13 handbooks for students, 86 patents, and reports on Russian and international scientific conferences.

Vinokurov Vladimir Arnoldovich – was born in 1950, July 25 in Aniva town, Sakhalin region. In 1972, he graduated from chemical and technological faculty of I.M Gubkin Moscow Institute of Oil Chemical and Gas Industry. In 1976, he graduated as PhD with the work "Elecrophysical properties of composite materials based on polymers with conjugated bonds", and in 1982 obtained Doctor of Sciences degree on the theme "Novel methods of functionalization of organic compounds". Since 1991, he works as the head of department of physical and colloid chemistry in I.M Gubkin Russian State University. He is the real member of Russian Academy of Natural Sciences. His scientific interests consist of the spheres of organic synthesis and catalysis, extraction under super critical conditions, additives to petroleum. Under his name over 150 scientific works were published, including 8 handbooks for students.

Kolesnikov Sergei Ivanovich – was born in 1959, June 12 in Moscow. He entered I.M. Gubkin Moscow Institute of Oil Chemical and Gas Industry in 1976, from which he graduated in 1981 by the specialty "Technology of oil and gas". After that, he worked as a junior researcher in the department of oil and gas technology. In 1989, he graduated as PhD on the theme "Calorimetric study of phase and polymorphous transactions in oily systems". Since 1992, he works as a senior researcher in the department of physical and colloid chemistry. Since 1998, he is the head of laboratory of industrial kinetics and catalysis. His scientific interests are: catalysis and catalytic processes, additives to petroleum, ecology of wastewater, and thermodynamics. Under his name more than 60 scientific works were published, including 2 monographs and 8 patents.

Chapter 1

MAIN TERMS IN THERMODYNAMICS

1.1. INTRODUCTION

1. In thermodynamics of physicochemical processes two methods of their study are separated — the study of equilibrium and non–equilibrium processes. The equilibrium ones are the processes proceeding at an infinitely low rate between two states of the working substance at a short distance from an equilibrium state of the system under the condition that no end changes happen in the surrounding. Equilibrium thermodynamics include the list of laws, principles and roles, which enable accurate calculation of the initial and end states of the present system to be conducted. It does not take into account intermediate states of the system and types of processes proceeding during consecutive transition from one state of the system to another at a finite rate of change of working substance states.

To solve applied problems with the help of methods of the equilibrium thermodynamics it is unnecessary to know the structure and composition of intermediate stages (under equilibrium changing of the composition and state of the system). The equilibrium thermodynamics enables material and heat balances, efficiencies of heat machines, equilibriums of physical (phase transformations) and chemical (equilibrium composition) of the processes, electrochemical processes and energy changing (of chemical affinity) of chemical reactions to be calculated.

2. Thermodynamics of irreversible processes studies regularities of physico–chemical processes in time. According to R. Clausius, classic thermodynamics mostly studies one–way spontaneous processes and does not consider non–spontaneous ones. Spontaneous processes proceed to the side of reduction of the free energy excess in the system with its transformation into linked heat. Thermodynamics of irreversible processes, grounded on R. Clausius theorem (1850), had been initially included into the classic thermodynamics, and then, with the help of postulates and theorems by I.R. Prigozhin *et al.*, was formed as a separate branch of thermodynamics.

Of great attention in thermodynamics of irreversible processes are sources of entropy production and regularities of entropy growth with time, which is bound to consideration of spontaneous processes only. Of insufficient attention is changing of free energies with time. Not attention is paid to non–spontaneous processes.

To consider thermodynamics, generally, the method of parametric processes (by Yu.S. Cherkinsky) can be applied, which broadens the general phenomenological approach to the study of thermodynamic acts.

In thermodynamics intensive and extensive parameters are selected, which can be subdivided into independent (parameters) and dependent (functions) variables. Yu.S. Cherkinsky adds thermodynamic theorems of irreversible processes by the principle of superposition.

The principle of superposition defines that each generalized parameter in any systems responds to the same parameter, but with the opposite sign. This is displayed most clearly in interactions of open and closed systems with the surrounding, which is determined by the Equation:

$$a' \cdot P_i' = -a' \cdot P_i'', \tag{1.1}$$

where a is a kinetic index, which reflects any property of the flow;

P_i', P_i'' are parameters acting in the present system and the surrounding counteracting to them.

The principle of superposition also determines interaction between spontaneous and non–spontaneous processes (of the system and the surrounding).

This principle is the broadened interpretation of the third Newton law. However, it also displays interaction and counteraction of scalar and vector, homogeneous and heterogeneous parameters (the Onzager principle) in the systems, homogeneous or heterogeneous by their phase and chemical composition.

Action and counteraction can be displayed between parts in a single system or between systems and the surrounding.

Action can be considered as proceeding of a spontaneous process, and counteraction — of non–spontaneous one — of force or energetic type and vice versa. According to Cherkinsky, the following equation is preserved for closed and open systems:

$$\sum P_i' + \sum P_i'' = \text{Const} \tag{1.2}$$

3. Hence the essence of classic and parametric thermodynamics and thermodynamics of irreversible processes was defined above.

From these definitions it can be noted that they do not reflect participation of spontaneous and non-spontaneous processes in general ones, and structure and nature of them. Further development of thermodynamics of irreversible processes must be grounded on the formulation of new additional principles and theorems, which take into account the interaction of above-mentioned pairs.

All processes must be considered from grounds of the interaction of spontaneous and non-spontaneous processes, their structure and nature. Division of equilibrium proceeding processes into spontaneous and non-spontaneous ones is untrue, because the system is so close to the equilibrium on the present stage of proceeding that type of the process can be hardly determined in this case and spontaneous can be hardly separated from non-spontaneous ones.

1.2. Terms of Systems and Processes

1. The system represents one of the main terms in any thermodynamics, for which various mathematical models and dependencies, models of their interaction and development (positive or negative) are created.

By a system is meant a part of space (real or logically thought out), separated from the environment by an interface (real or hypothetical) and possessing a definite composition, state and structure. In this case, it is meant that the system by definition is of the finite size, because no parameters, their properties, structure and corresponding terms are not thought out yet for the infinity.

The system composition is reflected by totality of atoms, molecules, more complex structures (fractals, micelles, floccules, phases) interacting with each other. The exchange of energy and substance proceeds between systems and the environment. The same can happen between parts of the system. State of the system is defined by values of intensive and extensive parameters. A discrete type of the substance composition defines structure of the system.

Recently, it is suggested that external fields or the absolute vacuum are also characterized by the discrete structure (Steven Hocking), in which, however, continuous and interconnected processes of the field transformation into a particle and, vice versa, a particle into field, i.e. interconnected spontaneous and non-spontaneous processes.

Macroscopically, structure of the system is also defined by regularities and rules of crystallochemistry, and polyhedron is the elementary spatial group of atoms (ions) of structure of the solid. Microscopically, the smallest constituents of the absolute vacuum structure are quark and attendant particles.

2. Systems are subdivided into three main types: isolated (hypothetical), closed and open. The isolated system, as the hypothetical one, does exchange neither energy, nor substance with the environment. All processes can proceed only in the current system, and finally the system is reduced to the equilibrium state by a series of fluctuations. However, equilibrium change of the working substance state cannot be performed in such system. All processes are non-equilibrium in the absence of compensation. Such hypothesis is suitable for studying processes in thermodynamics of equilibrium systems and for separating equilibrium systems from non-equilibrium ones, and equilibrium processes from non-equilibrium ones. Appropriate criteria for determining the state and direction of processes' proceeding were created for the isolated systems. Thermodynamic equations in an implicit form are easily deduced for these abstractions. Analytical analysis of equilibrium processes does not take into account their unsteady-state proceeding and influence of temporal factors on them.

Closed are systems which can exchange energy with the environment and is unable to exchange substance. Closed systems possess a cover impermeable for the substance, for example, glassy, metal, or other. But the cover substance contacting with the environment transfers energy to it as heat, radiation, etc.

Energy can transit (the energy excess) from the system to the environment or, at lower amount of energy in the system comparing with the environment, the system can accept the energy excess from it up to the level, at which intensive parameters of the system and the environment become equal. Such abstraction complicates description of the system behavior and the environment, and the analytical connection between different events.

Open are systems exchanging substance and energy with the environment. Open systems possess diaphragms semi-permeable for substance, and they contact with the environment through the interface. Mathematical description of these systems is grounded on methods of thermodynamics of irreversible, spontaneous and non-spontaneous processes, divided into steady state and non-steady state ones.

3. Processes are any changes in the system, including changes of its state, composition and structure under the effect of internal and external parameters of the medium. The process nature is characterized by appropriate form of the substance motion: physical, chemical, mechanical and biological, as well as combinations of these. All these processes are linked with each other, the lowest form of motion being a part of the higher one.

Processes are subdivided into thermodynamically reversible and irreversible, steady state and non-steady state, spontaneous and non-spontaneous ones. Quasi-equilibrium processes are nearly equilibrium ones.

Reversible processes, as the border case of quasi-equilibrium ones, are a suitable abstraction simplifying the view of thermodynamic equations and the structure of equilibrium thermodynamics. These processes form the basis of classic, equilibrium thermodynamics. Reversible are the processes, which are infinitely slow during infinitely long time and pass through the same equilibrium states during proceeding (points on the equilibrium curve of the process state change). Such processes cause no changes in the environment, and changing of the state of such system depends on the initial and end state of it and is independent on the way of the process proceeding. Time, as a parameter, does not participate in the thermodynamics of equilibrium processes.

As mentioned above, reversible processes represent a suitable abstraction, and equations of such processes are border theorems of real ones.

The processes are irreversible, when proceeding at a finite rate between two states of the system. Time, as a parameter of the process, plays a definite role in them. In particular, changing of the system state with time determines non-stationary conditions of the process proceeding. Irreversible processes are characterized by a mechanism of changing of the system state, and the function of processes depends on the way of the processes proceeding. If the working substance state in direct and opposite way changes irreversibly in terms of thermodynamics, the ways of changing of the system state in these directions are different. This characterizes losses of the efficient energy at its transformation into a useless one and reduces the efficiency of usage of the process energy.

4. Non-steady-state processes reflect their proceeding with time at the rate rise or decrease. These processes determine excessive accumulation of substance or energy in the system during a definite time, excessive expense of substance and energy from the system in accordance with the law of conservation of energy. Non-steady-state processes are defined as transitional ones, reflecting a change of the system from the initial to the end state (steady state) during a finite time. Non-steady-state processes are the most complicated real processes and can be described by a combination of complex integral-differential equations only applied to the vector mathematics.

Non-stationary conditions of proceeding of physicochemical and other processes are determined, when the first derivative of a process parameter by time is not equal to zero:

$$a \cdot \frac{\partial P_i}{\partial \tau} \neq 0, \tag{1.3}$$

where P_i matches concentration, temperature, chemical variable and other particular variables, e.g. it can be an intensive or extensive parameter (internal energy, entropy, etc.); a is a coefficient.

5. In the case, if physical and physicochemical processes in this particular system interacting with the environment proceed irreversible in thermodynamics, but at a constant rate (changing of the parameter P does not depend on time τ), such processes are names the steady state ones (for example, the Bodenstein steady state principle was formulated for chemical processes). Steady state processes are characterized by equality to zero of the first derivative by time:

$$a \cdot \frac{\partial P_i}{\partial \tau} = 0. \tag{1.4}$$

Various industrial processes proceed under stationary conditions (fractionation of oils, catalytic processes in a flow of reagents, heat-exchange devices). Biological systems also work under stationary conditions, although this term can be applied to a limited period of time and not to the whole organism, but just to some separate processes – for example, Crebbs cycles, etc. Under real conditions biological organisms continuously change with time, i.e. are under non-stationary conditions. Generally, they are spontaneously developing and oppressing.

Contrary to transitional non-steady-state processes, the steady-state ones can be described by simple mathematical equations. Such processes are regulated by various parameters under external influence on the working substance for the purpose of keeping the appropriate state of the system.

6. One of the most important terms in thermodynamics of irreversible processes is spontaneous and non-spontaneous proceeding of the processes (induced by external or internal forces to proceed towards accumulation of the energy excess in the system, i.e. a store of potential energy). The processes are spontaneous, when the system inevitably eliminates the free energy excess, accumulated in it. Comparing with the energy of the environment or other systems, this energy excess can be effectively transformed into the efficiency or uselessly transform into heat, scattered in the environment at constant temperature.

The free energy excess in the current system can be accumulated in consequence of continuous redistribution of energy and efficiency of the surrounding between the systems and the environment. One of the systems transfers the free energy excess to the environment. Another system absorbs from the environment and accumulates it for further loss of this energy excess at the transition to a lower energy level or during conduction of work under interaction of the working system with the environment, and so on. Such exchanges of energy and work between the environment and systems in the Universe are continuously proceeding and interrelated. The number of such systems and processes in the Universe, which cause exchanges of energy, substance and work, cannot be counted. The exchange of energy, substance and work between systems in the surrounding or these and the environment in the Universe is the integral property of the nature. Spontaneous processes are the source of excessive energy. In their turn, non-spontaneous ones are consumers of accumulated energy excess in a conjugated system to transform it into work or other forms of energy.

That is why, from grounds of thermodynamics of spontaneous and non-spontaneous processes, it is absolutely senseless to talk about beginning and end of the matter, about beginning and end of the Universe existence. The world around us and we ourselves in it, as a particle of the Universe, are the property of the matter display, the property of realization of the absolute vacuum and its development to one or another side as a result of interactions between spontaneous and non-spontaneous processes. From these grounds, the idea of a possibility to create the matter, worlds, and the Universe by other reasons, except the matter itself and its exhibition, is scientifically senseless.

Processes are non-spontaneous, which can proceed at the expense of work (mechanical, physical, chemical, electrochemical, etc.) or energy feeding lower power-consuming system from higher one. Consequently, it follows from this definition that non-spontaneous processes cannot be separated from spontaneous ones. In nature and technique these processes proceed simultaneously and continuously, are mutually independent and can proceed at the sacrifice of the above-mentioned property of systems.

In conclusion, it should be noted that at the present time thermodynamics of irreversible processes is generally developed for irreversibly proceeding spontaneous processes neglecting their interconnection with non-spontaneous ones. This book suggests consideration of thermodynamics of irreversible processes in interrelation to spontaneous and non-spontaneous processes.

1.3. THERMODYNAMIC PARAMETERS AND FUNCTIONS

1. In thermodynamics (general, technical, physical, field, irreversible processes, etc.), variables (parameters) are subdivided into independent and dependent, intensive and extensive, thermal, mechanical and calorimetric ones. Independent parameters include basic simple and composed complex ones. Basic parameters are those, which determine a single property of the matter and substance. They are, for example: time - τ [t], length – l [L], mass – m [M] and temperature – T [T]. Letters in brackets show dimensions of basic parameters. Theoretically, internal content and origin of these parameters are not established yet, and from the point of view of philosophy, the original cause grounding them is not determined. There are formal definitions of basic parameters of the following type: temperature, time, mass and length are measures of heating, existence, disposition and extent of an object. However, these terms can be hardly transferred to the absolute vacuum. Properties of the absolute vacuum are not studied yet, methods of measurement conduction in the absolute vacuum and parameters, which can define changes in it, are not determined.

This concise information reflects incompleteness of the scientific and natural knowledge about properties, state and structure of the matter.

2. Composed parameters are obtained by uniting basic parameters for the purpose of separation of an appropriate property and state of the system – density, pressure, heat capacity, etc. The content of some composed variables can be illustrated. For example, to reflect impulse of the substance particle effect in the present system the "pressure" term is used:

$$P = \sum f_i \tau / t, [M][L^{-2}],$$

(where τ is the duration of the force effect on the surface; t is time of the force effect).

The measure of energy fluctuations in the substance state of the system is heat capacity c_p, J/mol·K, the measure of the substance mass inclusion into a closed volume – density $\rho = \frac{m}{v}[M]\cdot[L^{-3}]$. Composed parameters represent a suitable method for studying properties, structure, symmetry and state of the system.

3. Thermodynamic functions are parameters reflecting dependency of the current parameter on basic and composed variables. Examples of thermodynamic functions are internal energy U, enthalpy H, entropy S, Helmholtz energy A and Gibbs energy G, Massier function A/T and Plunk function G/T. Thermodynamic functions display the following properties:

- Additivity;
- Differential of the thermodynamic function is full;

- Thermodynamic functions become simplest, if depend on natural variables; transition from one function to another can be smoothly performed with the help of Maxwell's transformations connecting basic and composed variables.

Consequently, thermodynamic parameters are subdivided into independent and dependent ones. In accordance with the set task variables can exchange their places.

4. Thermodynamic parameters are subdivided into intensive and extensive ones. Extensive parameters admit summing up of parts to obtain the unity, and vice versa. For example, total energy of the system includes the sum of potential and kinetic energies:

$$E = P + K. \tag{1.5}$$

The internal energy of the system equals to the sum of internal energies of system parts:

$$U = \sum U_i . \tag{1.6}$$

According to the Dalton and Amage law (ideal laws)

$$P = \sum P_i , \tag{1.7}$$

$$V = \sum V_i . \tag{1.8}$$

Pressure and volume equal to the sum of partial pressures and volumes of ideal gases of different nature, respectively.

Intensive parameters possess no additivity. They reflect potential tension of the system and are able just to be adjusted, when change in a definite direction. Intensive parameters are temperature T, pressure P, chemical potential μ, partial molar values $\overline{U}, \overline{H}, \overline{S}, \overline{A}, \overline{G}$, and others, surface tension σ, electric field intensity E_i, magnetic field intensity E_H, interaction energy between ligands and cations in polyhedron ε, etc.

5. Separation of thermodynamic processes into intensive and extensive ones enables structure of equations to be determined for calculation of various types of energy and work (heat).

Work and heat are special methods of energy exchange between substances. The work is the amount of energy transferred from one substance to another with the change of external parameters, and without change – the heat. Work and heat are marked as W and Q. Work and heat differ by the way of energy transformation. For example, work can be directly transformed into a store of excess of any energy in the system (electric, magnetic, potential, etc.).

Heat, without preliminary transformation into work in a thermal machine, cannot be the source of any other energy, it can be only spent for growth of the internal energy of the system:

$$Q = \Delta U. \tag{1.9}$$

It is customary to assume that:

$+W$ is the work expensed for change of the environment state;
$-W$ is the work loading the system;
$+Q$ is the heat loading the system;
$-Q$ is the heat transferred from the system to the environment;
$+\Delta U$ is growth of the internal energy;
$-\Delta U$ is diminution of the internal energy.

Heat, work and change of the internal energy are interconnected with the help of an analytic expression of the first law of thermodynamics. Any work can be expressed via multiplication of intensive X_i and extensive Y_i parameters:

$$dW = \sum X_i dY_i . \tag{1.10}$$

Definitions of work and heat already contain thermodynamic elements of spontaneous and non-spontaneous processes. For example, heat can be transformed into work at simultaneous proceeding of spontaneous and non-spontaneous processes in a thermal machine and, vice versa, work transforms into heat. At the gas heating in a cylinder under a plunger (non-spontaneous process), the gas expands (spontaneous process) that provides plunger rise (non-spontaneous process), and so on.

6. The notion of parameters and functions was formulated for equilibrium systems or equilibrium proceeding processes. These notions are also kept for equilibrium proceeding processes, but quantitatively parameters are continuously changed during non-stationary processes, and determination of numerical value of a parameter at every current moment of time requires a clear dependency of the parameter on time in appropriate coordinates:

$$X_i = X_i\left(P_i, \tau\right), \tag{1.11}$$

$$Y_i = Y_i\left(P_i', \tau\right), \tag{1.12}$$

where P_i', P_i are spatial coordinates; τ is time.

1.4. Classification of the Matter Objects and Forms of Motion

1. A matter of inquiry of natural and scientific disciplines is the matter in the form of substance or field. Changes of state of the matter and its field or substantial display with time can be studied in thermodynamics of spontaneous and non-spontaneous processes (in addition to other natural sciences). This study of processes and systems including them makes suitable leveling of the material matter according to the method of its complication (complication of particles in composition and structure). Somewhat classification of the world can be presented as a diagram displayed in Figure 1.1. Each level, which characterizes properties and state of the matter and its substantial view, can be described by appropriate selection of extensive and intensive parameters. Methods of parameters change and nature of parameters are reflected in names of thermodynamics – equilibrium thermodynamics and thermodynamics of irreversible processes, thermodynamics of coordinated systems and surface phenomena, thermodynamics of spontaneous and non-spontaneous processes, technical and physical thermodynamics, etc.

Each of these directions uses general laws and rules of studying physical, chemical, physicochemical and other systems and processes.

2. The following parameters are used in thermodynamics as extensive ones: internal energy U, enthalpy H, entropy S, Helmholtz energy A, Gibbs energy G. The internal energy and enthalpy are based on the first law of thermodynmics; entropy, Helmholtz and Gibbs energy are based on the first and the second laws of thermodynamics (Clausius, Helmholtz, Gibbs). As extensive parameters are also used square of the working substance surface, coordination numbers, etc.

From thermodynamic grounds, scantily investigated sphere is absolute vacuum and transitions of types: absolute vacuum-quarks, absolute vacuum-electromagnetic, gravitational and other fields. Formation /birth, delivery/ of quarks is a non-spontaneous process, and their transition into the absolute field is spontaneous, but the opposite is not excluded. This means that birth of quarks is a spontaneous process, and transition of quarks to a field of definite nature requires energy expense.

3. Spontaneous processes are the source of energy for the work conduction, and non-spontaneous processes proceed at the work expense or use the energy excess for changing the system state or production of a new property of energy or work. Work or energy can be analytically presented as a product of intensive and extensive parameters. Generalized parameters were introduced according to Yu.S. Cherkinsky: P_i – extensive parameters, and X_i – intensive parameters. Then work, heat or any other type of energy can be presented in the form of integral or differential equations as follows:

$$Э = P_i \cdot X_i, \tag{1.13}$$

$$dЭ = X_i dP_i + P_i dX_i, \tag{1.14}$$

where $Э = U, H, S, A, G$, and $X_i = P, T, \overline{V}, \gamma, \sigma$, etc.

Thus, it should be noted that the work term could only be applied to any system. Other parameters are applicable for macrosystems only.

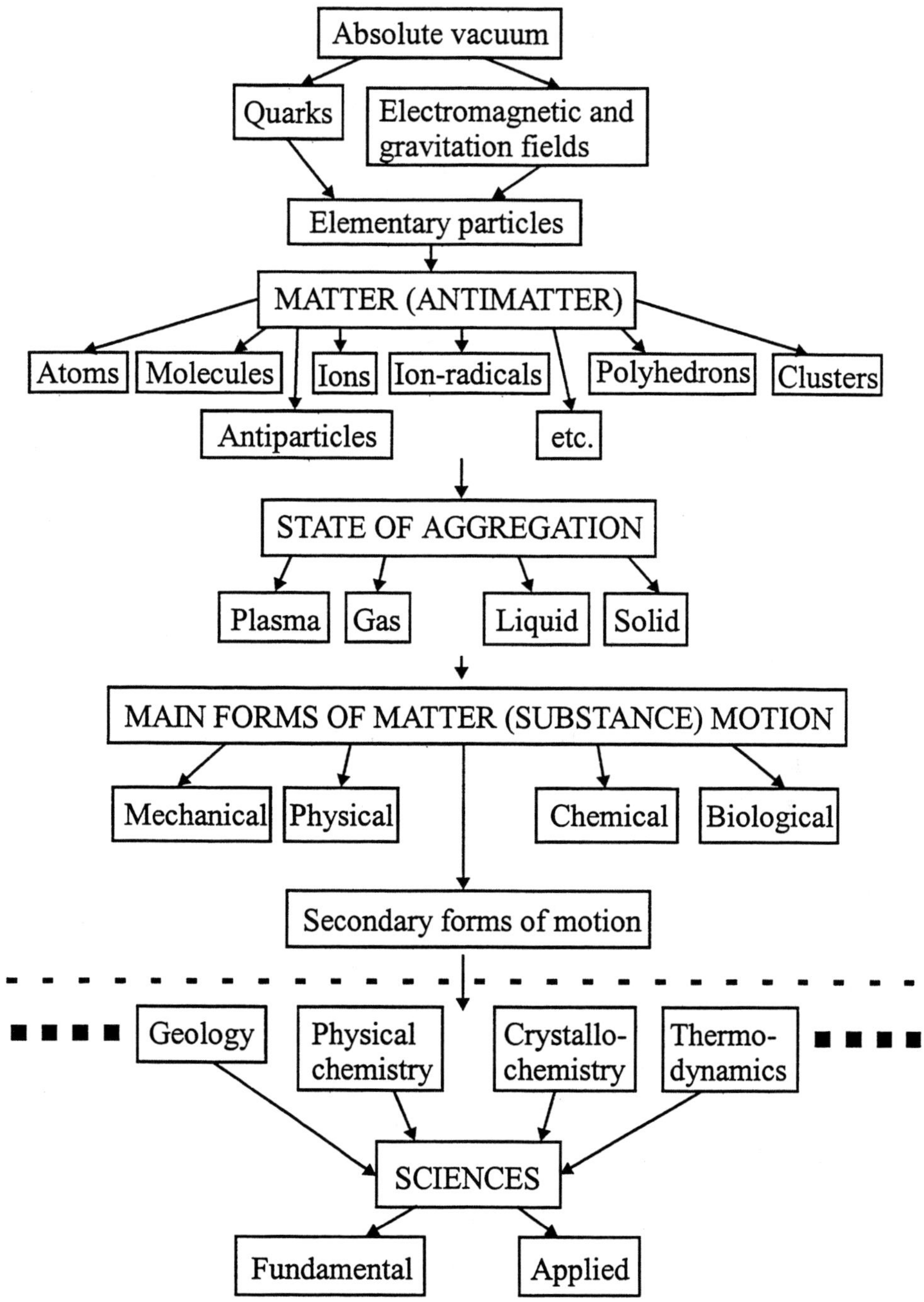

Figure 1.1. Classification scheme for levels of the material-substance state of the world, its development and degeneration (classification of sciences is shown below the dashed line).

1.5. TIME PARAMETER IN THERMODYNAMICS

1. Time is the parameter that defines regularities of proceeding of steady state and non-steady state processes in the phase-open systems and the environment. Directivity of time flowing in nature is undefined, and time itself in nature can be considered as a dimensionless parameter. In local objects or historically finiteness, time of beginning and end of a process can be marked. For these local objects, this allows the term of rate of process proceeding to be introduced. The nature is indifferent to division of processes to spontaneous and non-spontaneous ones. It can be noted only that such processes proceed continuously in nature, interrelated and equally probable. The latter means that each spontaneous process complies with a definite non-spontaneous one; they play the role of accelerator and decelerator of each other. Processes in nature and technique proceed so that spontaneous ones stimulate proceeding of non-spontaneous processes, and the latter raise stores of energy and substance in interacting systems, which, in their turn, can be the source of spontaneous processes overwhelming the effect of non-spontaneous ones, an so on.

Time as a parameter participates in equations, which reflect proceeding of spontaneous and non-spontaneous processes. Equations possess equal structure, but differ by sign of appropriate terms of equation (these terms can mathematically determine such processes as convection, diffusion, mass and heat exchange, kinetics of physical and chemical processes, hydrodynamics of flows, and other phenomena). Usually, time participates in equations of balances in the first degree. Rise of degree for such parameter as time complicates equations and make difficulties for interpretation of them.

2. Interconversions of energy and substance. Interconversions of energy and substance are analytically determined by the Einstein equation:

$$E = m \cdot c^2 \tag{1.15}$$

or by the Lorentz equation

$$m = \frac{m_0}{\sqrt{1-(\upsilon/c)^2}}, \tag{1.16}$$

$$\tau = \frac{\tau_0}{\sqrt{1-(\upsilon/c)^2}}, \tag{1.17}$$

where m is the mass of a particle; c is the speed of light; m_0 is the mass of quiescent particle (body); υ is the velocity of the particle (body) motion; τ_0 is time of reading for quiescent body (particle); τ is time of reading for moving body (particle). These formulae determine ability of a particle (body) to change mass and time of the mass change during its motion. This approach is formal and does not reflect mechanisms of elementary particle changes, which define the nature of time (tachione) with the particle (body) motion change.

These formulae are empirical. Actually, the velocity of motion of particles near black holes in the Universe can exceed the speed of light c. Then imaginary numbers appear in denominators of the above-mentioned formulae.

3. The time parameter is passive, because there are no methods to change it under usual conditions, no methods of any influence on time or change of it to any side do exist. Space-temporal change of these two parameters may probably happen near black holes, the gigantic space masses only. However, some phenomena do exist, which cannot be commonly understood. For example, conditions of the time fluctuation, regularities of its participation in changing state of the absolute vacuum and change of other phenomena are unknown.

Temperature and mass of a body represent active parameters, induced to change (heating or cooling), and mass transformation into energy can proceed in relation to the Einstein equation, etc. These parameters affect the state of the present system and the medium surrounding it. They are subjected to stratification and fluctuations, interact with each other causing cross phenomena /effects/ in nature (thermal diffusion, thermal electricity, etc.). According to the response to external actions basic parameters can be formed into a sequence:

$$T > M >> L >>> \ldots > \tau. \tag{1.18}$$

As this takes place, a single direction of change of the system structure was observed before – growth of entropy as a result of interaction of external and internal factors and change of the system state (spontaneous processes). Disposition of parameters in the above-mentioned sequence is conditional. Dots in it display a possibility of increase of the number of basic parameters suitable for studying phenomena under development of knowledge about the nature of processes.

Disclosing methods of external parameters effects (pressure, light energy, electromagnetic effects, etc.) on the basic parameters, the internal connection between different phenomena and processes is determined. Methods of the time parameter change under the effect of external and internal parameters of its extent, spreading in the space, and change of direction cannot be determined. Its homogeneity or heterogeneity cannot be determined also. That is why, such terms as time fluctuation of time localization, stretching or compression of time are not applied to the variable τ.

4. Time is an indirect parameter for such terms of the substance state change as the rate of its motion, mass loss, deformation or change of rates of these processes. The process rate is also passive parameter for the change of the system state: substance-environment. However, the process rate adequately responds to external effects and internal processes – mechanical and chemical influence. Generally, the rate determines a change of the parameter value during specific time, and acceleration – change of the rate during specific time. These parameters can characterize non-steady state or steady state development of processes in a phase-open system.

Non-steady state changes of the flow state of mass of substances, temperature and flow motion impulse can be presented by the following equations, which include convection component, mass and thermal transfer, diffusion and heat conductivity, chemical and physical changes:

$$-\frac{\partial c_i}{\partial \tau} = \mathrm{div}(c_i \cdot \upsilon) - D\nabla^2 c_i + \beta \cdot (c_{surf} - c_{vol}) + v_i \cdot w_i, \tag{1.19}$$

$$-\rho \cdot C_p \frac{\partial T}{\partial \tau} = \lambda \nabla^2 T - C_p \rho V \mathrm{div} T + \alpha (T_1 - T_2) + v_2 \Delta H_i \overline{w}_i, \tag{1.20}$$

$$-\rho_x \frac{\partial \upsilon_x}{\partial \tau} = -\rho_x \left(v_x \frac{\partial \upsilon_x}{\partial x} + \upsilon_y \frac{\partial \upsilon_x}{\partial y} + \upsilon_z \frac{\partial \upsilon_x}{\partial z} \right) + \mu \nabla^2 \upsilon_x + \gamma \Delta \upsilon_x + \frac{\partial P_x}{\partial x}, \tag{1.21}$$

or under condition that μ = const the equation of motion can be reduced to the following view:

$$\rho_x \frac{\partial \vec{\upsilon}}{\partial \tau} = -\rho (\vec{\upsilon} \,\mathrm{grad} \vec{\upsilon}) + \vec{g} \rho + \mathrm{grad} p - \mu \nabla^2 \vec{\upsilon}, \tag{1.22}$$

where c is concentration; τ is time; υ is the linear rate; D is the diffusion coefficient; β is the mass transfer coefficient; c_{surf} and c_{vol} are concentration of substance (on the surface) at the interface and in the substance volume, respectively; v_i are stoichiometric coefficients; x, y, z are system coordinates; ρ is the density; w_i the rate of chemical process; C_p is heat capacity; λ is the coefficient of heat conductivity; α is the heat emission (heat transfer); ΔH is heat of a chemical reaction; μ is the coefficient of dynamic viscosity; P is pressure; γ is the coefficient of motion transfer between phases.

These equations represent somewhat basis to describe thermodynamics and mass dynamics of irreversible processes, motive forces of which are convection, diffusion, chemical reaction, mass and heat transfer, and other factors affecting the flow of substances. Description of flows of electromagnetic and other types of energy, bound to fields, are of similar character.

Formally, equations (1.19) – (1.22) reflect the effect of densities, concentrations, temperature, and other parameters on the flow at some distance of flow cross-sections from each other. Detailed mechanism of the processes, which define the type of flow and excite it, can be created on the basis of unity of the methods of micro- and macroscopic description of systems and processes, proceeding in them on the atomic, ionic, molecular and other levels, and with participation of fields of different nature.

1.6. The Law of Conservation in Thermodynamics of Irreversible Processes

1. The law of conservation is the absolute law of nature and determines a store of matter and energy in the Universe, as well as their preservation under various types of change of the state and type of energy and matter. This law characterizes indestructibility of motion (energy as a measure of motion) and the matter, and determines a possibility of the energy transfer from one type to another at the sacrifice of participation of appropriate motion forms in the processes. Analytically, the law of conservation can be presented in the form of constancy of sums of matter and antimatter masses, energies and entropies with different signs:

$$\sum m_i = \pm\infty(\text{const}), \tag{1.23}$$

$$\sum \Delta U_i = \pm\infty(\text{const}), \tag{1.24}$$

$$\sum S_i = \pm\infty(\text{const}). \tag{1.25}$$

The sum of masses of substances and antisubstances, matter and antimatter in the Universe is constant. The positive sign is used for usual substances and matter, the negative sign – for antisubstances and antimatter. In line with discovery of absolute temperatures (I.P. Bazarov) in the negative area ($T = -\infty, 0, +\infty$), total store of the internal energy in the Universe is also constant, i.e.:

$$U = \text{const and } -U = \text{const or } \pm U = \text{const}.$$

The store of entropy in the Universe is also constant, though local changes of it are continuous and different by signs. But this does not define positions of L. Boltzman about existence of the primary equilibrium of entropy distribution in the Universe.

The law of conservation in the above-mentioned general formulation is scalar and possesses no gradients. The law of conservation does not depend on the equilibrium of entropy distribution in the Universe.

2. In a limited, local space, the law of conservation is easily spread upon processes of interconversion of substances and masses, energies and pulses of the motion amount. Particular consequence of the law of conservation is the first law of thermodynamics and the law of equivalence of interconversions of different types of energy into each other.

This law determines two particular expressions of the first law of thermodynamics:

$$W = Q, \tag{1.26}$$

$$Q = W. \tag{1.27}$$

Simplicity of these equations is connected to measuring of both parameters in J/mol units, which excludes necessity to introduce transfer coefficients into the equation.

3. The first law of thermodynamics and the law of equivalence in a usual form do not reflect existence of spontaneous and non-spontaneous processes in nature and technique. Spontaneous processes (decelerated by conjugated non-spontaneous processes) proceed in the definite systems at the sacrifice of free energy in them, i.e. excessive values of ΔU, ΔH, ΔA or ΔG. Systems with the minimum entropy can also be the source of spontaneous processes. At the same time, energy or work of spontaneous processes is the source of proceeding of non-spontaneous processes. As a result of conjugated interaction between spontaneous and non-spontaneous processes and vice versa, a change of the state of both the system and the environment happen.

In the Universe, it is impossible to name the original cause of occurrence of conjugated spontaneous and non-spontaneous processes. This is simply the state of the Universe, and the hypothesis of a great explosion as the reason of occurrence of spontaneous processes (spontaneous compression of the Universe and spontaneous explosion), should not be revised, because the nature of non-spontaneous processes is not discussed in this case. Incidentally, each spontaneous process is conjugately related to a non-spontaneous one, as an integral property of the nature.

Basing on the law of conservation several postulate notes can be put forward:

- An infinite number of pairs of conjugated spontaneous and non-spontaneous processes proceed simultaneous in the nature;
- There is no vector direction of the interaction of conjugated pairs of processes in the Universe;
- The Universe is indifferent to direction of its development. This property is adhered to it by a man basing on his finite vector development.

4. Several most typical examples of display of interaction between spontaneous and non-spontaneous processes in the range of one or several systems, able to exchange energy and substance, can be presented. Two systems of substances with different concentrations of the same substances in them, c_{A_i} and c'_{A_i}, can be chosen, and it can be accepted that

$$c_{A_i} > c'_{A_i}. \tag{1.28}$$

If P and T = const, this condition will be characterized by chemical potential of the first system being higher than that of substances of the second system, i.e.

$$\mu_i^1 > \mu_i^{11},\ \left(\frac{\partial G_i^1}{\partial n_i^1}\right)_{P,T,n_{j\neq i}} > \left(\frac{\partial G_i^{11}}{\partial n_i^{11}}\right)_{P,T,n_{j\neq i}}. \tag{1.29}$$

5. Then degradation will happen in the first system, and complication of the system – in the second one. A spontaneous outflow of substances proceeds in the first system, and non-spontaneous one – the process of induced feed maintenance of the second system by energy and substance from the first one – in the second system. As a consequence of two interacting systems, redistribution of substances will proceed up to equaling compositions of both systems, i.e. up to leveling of chemical potentials:

$$\mu_1^i = \mu_2^i. \tag{1.30}$$

6. This example clearly displays that both processes, spontaneous and non-spontaneous ones, cannot be separated from each other. For example, heating of a substance cannot be separated from its simultaneous cooling down, decrease of voltage in a conductor cannot be separated from its growth on its opposite end, dissolving – from precipitation, etc.

This problem will be discussed in more detail in the next Chapter.

1.7. Vector-Scalar Change of Energy

The surroundings can be presented as a totality of phase-open systems, differed by concentrations of substances and free energy value – its excess or deficit. Systems with excessive free energy strive to eliminate this excess, and systems with a deficit of free energy deliver it. Such redistribution of energy in the surrounding proceeds in accordance with the scalar-vector interpretation of direction of processes, their rates and splitting of free energy into components. The role of free energies under constancy of appropriate parameters in the present system and the conjugated one can be played by changes of Helmholtz and Gibbs energies. These energies possess a definite direction of redistribution and are characterized as vector parameters. These functions include internal energy, enthalpy and linked energy, which are characterized as scalar parameters of change of energy of systems. Conditionally, they can be graphically depicted, as shown in Figure 1.2.

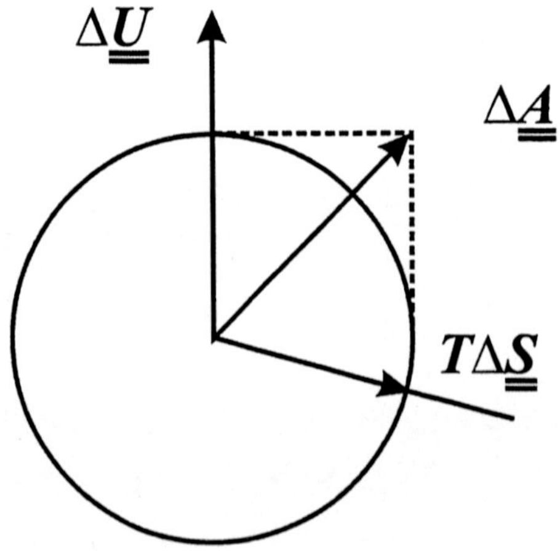

Figure 1.2. Vector-scalar display of the Helmholtz energy for the present system at $T = \text{const}$.

This presentation reflects possibility of a local directed redistribution of the free energy in space, i.e. energy redistribution between two local objects.

Chapter 2

SPONTANEOUS AND NON-SPONTANEOUS PROCESSES

2.1. IDEA ABOUT SPONTANEOUS AND NON-SPONTANEOUS PROCESSES

1. Spontaneous and non-spontaneous processes can be easily separated in the nature. It can be noted more strictly that such processes proceed intimately with each other, and just conditionally it may be cited that in the present system a spontaneous process is preferable, and in the other system – non-spontaneous one under the effect of external factors.

Spontaneous processes happen in the case, when the store of free energy in the system exceeds the average energy store in the surroundings or a system contacting another one. In response to excess of free energy, the present system strives to spontaneous elimination of it in any way – by a change of temperature, pressure, redistribution of substance, etc. These processes proceed in phase-open systems in the absence of compensation at a finite rate and, consequently, by a non-equilibrium method, i.e. thermodynamically irreversibly.

Spontaneous processes proceed in one direction only, particularly, decreasing the free energy excess down to reaching the state of equilibrium in interacting systems. In any system, spontaneous processes proceed at a finite rate. This is determined by events decelerating the process, which relate to non-spontaneous processes. For example, cooling of a heated system is restrained by heating of the surrounding or interacting systems. Such mechanism of proceeding of two different processes promotes a slow approaching the state of the surroundings by the system or the state of substance interacting in the system.

2. Consequently, decelerating processes are related to reverse or non-spontaneous ones. They can proceed due to external feeding of energy (in this case heating of substances, radiation, ultra-violent radiation, etc. happen in the present system). Work feeding of a working substance or a system – gas compression, or energy transfer, substance and work transfer from one system to another are also non-spontaneous processes.

In the latter case, a spontaneous process stimulates proceeding of a non-spontaneous one, or to put it more strictly, in the latter case, a non-spontaneous process in another system (absorbing energy, substance or work) prevails over a spontaneous one, and the latter reduces the system into more regulated state and with higher store of energy, which exceeds the average energy of the surroundings or other systems.

3. Continuous redistribution of energy and substances between systems proceeds in the nature, which leads to decrease of the free energy amount in one system and rise of the store of it, substance and regularity of structure in another one.

In thermal machines based on simultaneous proceeding of spontaneous and non-spontaneous processes between three substances a work is produced and, vice versa, if a work feeds the working body, a cooler is cooled down to lower temperature and heating up of a heater. In a thermal machine, there proceeds a spontaneous transition of energy in the form of heat Q_1 from heater to the working substance and a non-spontaneous transfer of difference of heats $Q_1 - Q_2$ into work W, where Q_2 is the heat transited from the working substance to the cooler. Hence in a thermal machine (as well as in any other type of source) two processes, spontaneous and non-spontaneous, are united, which proceed simultaneously and continually. Heating of the cooler is a spontaneous process, and heat transfer from the working substance to the cooler is a non-spontaneous one. In nature, equilibrium processes are absent, and this term is useful for thermodynamics of equilibrium processes only. Equilibrium processes are not divided into spontaneous and non-spontaneous ones, because equilibrium processes display infinitely small deviations from the state of equilibrium, and at any moment of time any direction of them is equally probable.

2.2. Self-Organization as a Consequence of Influence of Non- Spontaneous Processes on Structure, Composition and State of the System

1. The phenomenon of self-organization in nature and technique is a usual state of a system, and processes of self-organization can be of thermal, mechanical, physical, chemical, biological or any other origin. At the present time, four main forms of the matter motion are separated: mechanical, physical, chemical, and biological. In nature and technique all these forms are mutually bound, and secondary forms of the matter motion are grounded on them, among which are mechanochemical, physicochemical, biophysical, biochemical, and many others.

2. Structuring in systems with prevailing of one of forms of the matter motion or combination of forms can proceed under the effect of energy and work (mechanical, electrical, etc.). As a consequence, symmetry, structure and state of a system or an object are improved or perturbed. Non-spontaneous processes proceeding in the system result in a self-development with complication of its structure, composition, and state.

Examples, several of which below, can be presented on each form of the matter (substance) motion, which reflect different levels of self-development of the system.

2.2.1. Self-Organization of System by Convection

1. Convection in liquids and gases is a result of partial heating of some and cooling of other layers of liquid and their mutual mixing. Heated up layers of liquid possess lower density comparing with cool ones. This causes downthrow of cool layers and rising of heated ones. To explain the mechanism of self-organization of string flows in these layers of liquids and gases under the effect of a non-spontaneous process (in nature, heating up of a liquids happens by virtue of absorption of sunlight beams, in laboratory – by virtue of heat transfer from a thermal device to the liquid) the Benar cell method can be used (G. Nikolis, I. Prigozhin) for the heat application to the liquid to heat it.

2. Benar cells occur in liquid at their heating up to an appropriate temperature. Liquid in the definite volume at a constant temperature exists in the state of equilibrium. If heating of the lower layer of liquid is low, heat will be transferred from the lower to the upper layer, and temperature, density, and pressure of the liquid will be inhomogeneous by layer height of the liquid. Gradual rise of temperature difference between upper and lower layers of the liquid ΔT up to some critical value T_{cr} causes a sudden beginning of motion of the liquid volume. Upward and downward flows appear in the liquid as the consequence of difference in density. Such flows are absent at small difference in temperatures due to stabilizing effect of the liquid viscosity and heat conductivity. Convection vortex type flows appear in liquids at temperature T_{cr}, formed in interacting cells, conditionally displayed in Figure 2.1.

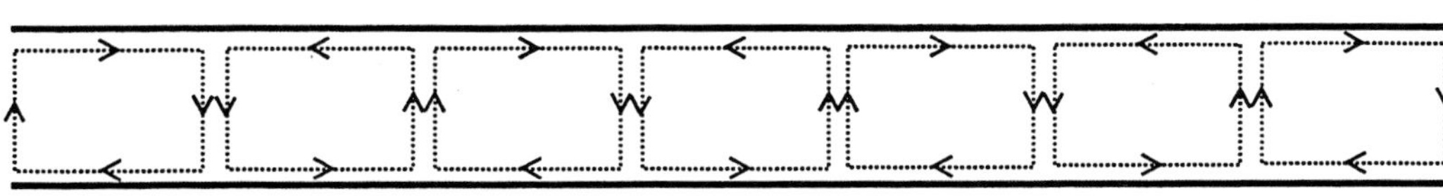

Figure 2.1. Benar cells and consistent interaction between them.

Such Benar cells already act consistently, they feel each other, and under this interaction the effect of non-spontaneous processes on organization of the system is displayed. In its turn, spontaneous processes (friction, energy and motion dissipation) perturb this organization. Benar cells represent already quite high level of self-organization of the system. Levorotary and dextrorotary motions of the liquid are separated in them, and such organization is always observed, when the critical difference in heating temperatures, ΔT_{cr}, is reached.

Rotation of the liquid in Benar cells under temperature difference ΔT_{cr} is quite predictable. However, rotation direction is of a random type. It is impossible to predict direction of liquid rotation at the current time. The former phenomenon is of the determinate type, and the latter is random, which defines ambiguity in self-organization of complex systems.

Further rise of temperature ΔT can enable the next critical temperature to be reached, at which the regime of liquid motion as cells transits into turbulent one reflecting higher order of organization of the system.

2.2.2. Self-Organization in Catalytic Processes

1. Catalytic processes proceed in a system containing a catalyst, initial substances and reaction products. In the absence of catalyst molecules of substances exist in a homogeneous state — energetically inhibited to transform into reaction products. This phenomenon reflects something like an initial state of liquid system before formation of Benar cells in it.

Benar cells occur in liquid under temperature effect, i.e. as a consequence of interaction between open systems — liquid and a heater.

2. Catalytic system is also open, and catalyst is an external parameter, which will define self-organization processes in the system with creation of cyclic ones of the Benar cell type in them. Non-spontaneous and spontaneous processes, proceeding continuously and depending on each other, can be separated from the cyclic ones. More distinctly notice that non-spontaneous process dominates over spontaneous one on a separate stage — with energy or work consumption, and on the other stage spontaneous process dominates over non-spontaneous one — with release of free energy excess or work conduction. Both types of the processes participate in self-organization of the system. In this case, non-spontaneous processes represents the basis of the system organization, improving its structural and energetic organization, and spontaneous one decelerates self-organization of the system as a consequence of entropy growth and reduction of a store of free energy.

3. The process of self-organization in a catalytic system is to create such structural and energetic composition from catalyst and reagents molecules, which enables processes of redistribution of bonds and electrons to be conducted in molecules with accompanied creation of a new substance with new molecules and other particles. In this case, each catalytic cycle can be conditionally determined as a Benar cell.

Catalytic process includes a significant number of spontaneous and non-spontaneous stages and sequentially passes through the following steps: diffusion of molecules to a catalytic center, absorption of a molecule on the catalytic center, energetically and structurally profitable orientation of the molecule in relation to the active center, activation of electrons on atomic and molecular orbitals, redistribution of electrons between activated molecules and bonds, some of which are obtained in the basic, i.e. inactivated, state, and others in activated one, transition of activated molecules into the main state, desorption of molecules from the active center, and diffusion of molecules into the volume.

4. This selection of stages, which describes the catalytic cycle, is continuously repeated with regeneration of the catalyst. A combination of conjugated spontaneous and non-spontaneous processes continuously acts in this catalytic cycle. They participate in both self-development of the system and its degradation. Self-development of the system cannot proceed without degradation of its, accompanying the former. This is the common property of the matter and its material and field states.

2.2.3. Deceleration of Spontaneous Process by Non-Spontaneous Ones

1. Spontaneous processes proceed at a definite rate by virtue of their deceleration by non-spontaneous ones and vice versa. This is confirmed by a finite rate of proceeding of any natural or technological processes: heating—cooling, diffusion—compression, chemical spontaneous and non-spontaneous transformation of substances, biochemical processes, convection transpositions of air flows, and others. These processes proceed during prescribed period of time. This is bound to appearance of tensile resistance to spontaneous or non-spontaneous proceeding of the process in any interacting systems.

This proceeding in the system can be clearly illustrated by the example of gas pressure redistribution between two communicating vessels (Figure 2.2).

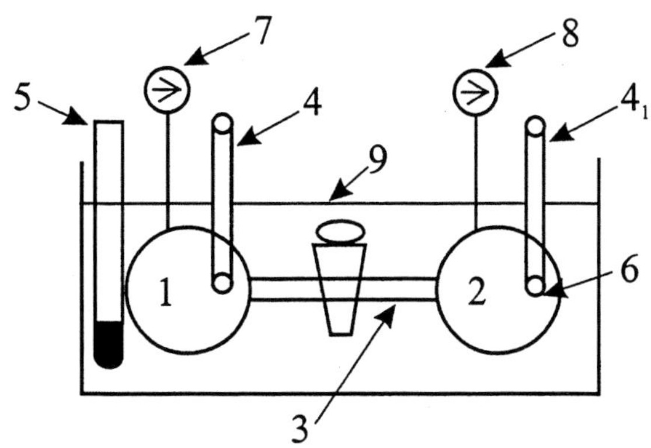

Figure 2.2. The scheme of device to illustrate pressure redistribution between two vessels.

2. A device can be imagined, shown in Figure 2.2, which is called the Gey-Lussac—Joule device. We will discuss the pressure redistribution between two vessels with a real gas, molecules of which interact with each other.

In the beginning the vessel 1 is filled with gas at pressure up to P_1 = 20-30 atm, and pressure in the vessel 2 is remained atmospheric, P_2 = 1 atm. The system is thermostatically controlled. Thereupon the valve 9 on the connecting pipe 3 is opened. The gas flows from the vessel 1 to the vessel 2. Pressure in the vessel 1 is reduced, and the gas occupies both vessels at the sacrifice of expansion from smaller volume to greater one (spontaneous process).

3. In accordance with the approximate Van-der-Waalse equation the value of the free energy for the real gas will decrease in the first vessel in respect to the equation

$$\Delta U = \frac{a}{V_1} - \frac{a}{V}. \tag{2.1}$$

Actually, reduction of pressure enables gas volume to be risen from V_1 to the total volume $V = V_1 + V_2$. Consequently, $V > V_1$ and the change of the internal energy with decrease, because

$$\frac{a}{V_1} - \frac{a}{V} > 0. \tag{2.2}$$

Simultaneous pressure boost in the vessel 2 from one atmosphere reaches $p_2 < P$. This is a non-spontaneous process. Consequently, reduction of the potential energy of the gas in the vessel 1 due to gas redistribution is consumed for growth of the potential energy in the vessel 2.

In this case, a fall of total pressure, in both vessels of $V = V_1 + V_2$, happens, and pressure in the vessel 2 boosts. The following inequalities will be valid under redistribution of the gas:

$$p_2 < P \text{ and } p_1 > P. \tag{2.3}$$

In the second case, which, of course, is considered together with the first one, energy is released at the sacrifice of validity of the inequality ($V_2 < V_1$):

$$\frac{a}{V_1} - \frac{a}{V_2} < 0. \tag{2.4}$$

Consequently, an excess of the internal energy is formed in the second vessel at the sacrifice of gas compression. This excess is released by the system into the surrounding, or at low rate of the heat release the system heats itself up. This experiment evidently displays the interaction of spontaneous and non-spontaneous processes in the system and with the surrounding.

In the absence of the work conducted or expended in the system (walls of the vessel are rigid), the first law of thermodynamics will be presented as follows:

$$Q'_{surround} + \Delta U'_1 = Q_{cool}, \tag{2.5}$$

$$\left(-Q''_{surround}\right) + \left(-\Delta U''_1\right) = Q_{heat}. \tag{2.6}$$

Heats of gas cooling in the vessel 1 and gas heating in the vessel 2 are not equal, which is displayed by inequality of temperatures, measured by thermometers 4 and 4_1. Manometers 7 and 8 measure reduction and boost of pressure in vessels 1 and 2.

In addition, notice also that the non-spontaneous process in the vessel 1 is gas cooling (below the equilibrium temperature), and heat transfer from the surrounding to the vessel

1 and gas in it is the spontaneous process. For the vessel 2, heat transfer into the surrounding is the non-spontaneous process, and heating of gas and the vessel 2 is the spontaneous one.

As yields from this experiment, spontaneous and non-spontaneous processes proceed continuously and are interconnected.

In nature, the example of such interacting systems are adjacent areas of high and low pressure in the atmosphere — cyclones and anticyclones, areas of formation and degeneration of clouds, various areas in oceans, etc.

Chapter 3

THE FIRST LAW OF THERMODYNAMICS FOR OPEN SYSTEMS

3.1. FORMULATION AND ANALYTICAL EXPRESSION

1. The first law of thermodynamics is based on the general law of nature – the law of conservation and, in particular, on the law of conservation of energy. This law is the absolute law of nature, which has no limits in any manifestations of the matter, its properties and composition. The law of conservation of energy determines that the total store of energy in the Universe is constant under any intertransformations of the matter and, in particular, in its material forms. Intertransformation of energy and substance happens in accordance with the law of equivalence. Transformation of some forms of energy into another is performed in the frames of four main forms of the matter motion: mechanical, physical, chemical and biological ones. These forms of motion are inseparable from the matter, material state of the world and are the main ones in their existence and change of the matter. The main forms of the matter motions are accompanied by the secondary ones.

Forms of the matter motion are bound to accompanying types of carriers for appropriate forms of motion, which determine their external manifestation, interaction and mutual influence. The lower forms of motion are a component of the higher ones, but not and not the reverse.

2. Perpetuity of existence and manifestation of the main and secondary forms of the matter motion meets difficulties from grounds of application of this notion to biological forms of motion. One can suggest that the Earth only is the carrier of biological forms of motion. However, the simplest carriers of biological form of the matter motion as aminoacids are spread in the outer space and can be found in meteorites, which have fallen down to the Earth's surface.

3. Open systems continuously exchange various forms of motion. The form of motion and its level depend on the self-development degree of the current system, disposition of systems in the Universe, and their lifetime. The exchange of forms of motion between different systems and inside them is bound to redistribution of energy and substance (masses) at the sacrifice of continuous, perpetuate and interstipulated proceeding of spontaneous and non-spontaneous processes.

4. Quantitative calculation of the energy redistribution between systems is performed for open (or closed) ones with the help of the first law of thermodynamics. In the usual formulation this law of thermodynamics allows no nature of proceeding processes to be cleared up, and spontaneous processes to be separated from non-spontaneous ones.

5. In a particular formulation, the first law of thermodynamics is presented by the following expressions:

$$dW = dQ, \tag{3.1}$$

$$dQ = dW, \tag{3.2}$$

where W and Q are work and heat, respectively.

According to equations (3.1) and (3.2) work can be the source of heat, or heat can be the source of work. Equation is valid for all processes. Work is intended to overcome friction, and friction heats up the working substance. Heat produced by friction scatters in the surrounding. Equation (3.2) determines that heat can also be the source of work. However, this condition can be fulfilled for systematic production of work in the presence of three substances in the system only – a heater, a working substance and a cooler. Simple heating of the working substance (a stone on the Earth's surface, for example) will not make it do make a work.

It is significant that equations (3.1) and (3.2) are grounded on the law of equivalence. Joule has determined that always when 427 kg/m of work is expensed, a single large calorie of energy as heat or 4.184 kJ of heat is released. There is no equivalence coefficient in equations (3.1.) and (3.2), because work and heat are measured in Joules, and, consequently, the transformation coefficient equals to 1.

3.2. Work and Heat in Spontaneous and Non-Spontaneous Processes

1. In thermodynamics, work and heat are defined as two types of processes of energy transfer between two substances. Work is the result of the force effect on the working substance, which enables an element or volume of the working substance to be changed. In the general case, pressure, surface tension, components of intensities of electric and magnetic fields, chemical potential, partial molar values, etc. can be the components of the force.

2. Heat is defined as an ability to redistribute energy between two interacting substances with rise of the Brownian motion effect, that is, it is bound to rise of disorganization in motion of particles, generally, in the united system. Heat can be bound to redistribution of the internal energy and enthalpy between substances at the sacrifice of molecular composition of the working substance, a change of osculation, rotational and other types of motion of particles in the system, etc.

These two processes are interconnected by the law of equivalence. Their values depend on the process nature and, consequently, on the way of proceeding of the processes, bound to heat and work. Heat and work cannot be accumulated in a system and stored until their further expense, because they manifest themselves at the very moment of their production or consumption. As shown below, a system is able to accumulate and store the internal energy or an amount of potential energy only.

3. Heat and work manifest themselves at the interaction of minimally two substances only. One of these substances is the source of force or store of the energy excess comparing with the working substance energy, and the second one is the instrument for conducting the work of substance or heat transfer. In its turn, the first substance must preliminarily interact with another one (the surrounding) in order to obtain an excess of energy and, consequently, a store of force able to deform the second substance – the working one, which under definite conditions can conduct a work or consume it.

4. In these interacting substances, some processes are spontaneous, and others are non-spontaneous ones. Processes bound to production of work and heat will proceed just under interaction of spontaneous and non-spontaneous processes. Number of these processes – spontaneous and non-spontaneous – in the surrounding is infinite, but these are precisely the processes which reflect development and degeneration of the matter, especially in its material state and existence.

5. Heat and work can load the system. Such processes are determined as non-spontaneous ones. Other systems contacting the current one will be sources of excessive energy or force and will affect the initial one.

System itself can produce work and release heat into the surrounding or other systems. These processes can be named spontaneous. Rates of proceeding of the mentioned processes and the level of consumption of their energy will be determined by decelerating reasons, based on non-spontaneous processes.

Analytical equations connecting work and heat of spontaneous and non-spontaneous processes are the following:

$$W_{n\text{-}sp} = Q_{sp}, \tag{3.3}$$

$$Q_{n\text{-}sp} = W_{sp}, \tag{3.4}$$

where *n-sp* and *sp* indexes mark non-spontaneous and spontaneous processes, respectively. The present equations display these two opposite processes to be continuous, inseparable and interconnected ones. Summing up these two expressions, we obtain:

$$W_{n\text{-}sp} + Q_{n\text{-}sp} = Q_{sp} + W_{sp}. \tag{3.5}$$

In the first expression, the system eliminates heat, and in the second – consumes heat and under definite conditions may conduct a single work (for example, gas expansion in a cylinder with a plunger).

After regrouping, it will be obtained:

$$W_{n\text{-}sp} - Q_{sp} = Q_{n\text{-}sp} - W_{sp}. \tag{3.6}$$

6. In open system (interacting with the surrounding), reasoning from equation (3.6), either storing of an excess of energy in it or consumption of the energy excess can be noted. This is determined by ratios of values in the left and the right parts of equation (3.6). If the state of working substance in the open system does not change at the sacrifice of thermal and mechanical processes, equation (3.6) can be equaled to zero:

$$\sum W - \sum Q = 0, \ \oint dW - \oint dQ = 0. \tag{3.7}$$

This condition is valid for circle processes. For non-circle processes, the difference between heats and works will equal not to zero, but to some value, usually marked by letter U and is call the internal energy.

3.3. Internal Energy

1. Internal energy is a new function, which differs by nature from work and heat. Work and heat are the functions of way, by which the process proceeds. Internal energy does not depend on the way of change of its value, but depends on the initial and end states of the system only. It is the function of a point.

For non-circle processes equation (3.7) in the differential form can be presented using the internal energy function:

$$dU = dQ - dW. \tag{3.8}$$

2. For open systems, change of the internal energy is connected with energy and substance exchange of the current system with the surrounding. Spontaneous and non-spontaneous processes are continuous and interconnected in them. Analytically, the exchange of energy of the current system with the surrounding can be presented in the form of equation:

$$dU = d_iU_{sp} + d_iU_{n\text{-}sp} + d_eU_{sp} + d_eU_{n\text{-}sp}, \tag{3.9}$$

where i and e indexes mark internal and external energy exchange, respectively; sp and n–sp are spontaneous and non-spontaneous processes, respectively.

3. In open systems the energy exchange depends on time and in some specific cases can proceed under stationary conditions. The Equilibrium energy exchange in real open systems is of idealized presentation. This change of the processes state cannot be obtained under real conditions and is the border change of state of the system interacting with the surrounding.

In the border state, differential of the internal energy is total, and the internal energy is continuous, one-valued and the finite function of state of the system; the change of U with parameters of the system depends not on the way of the process proceeding, but on initial and end states of the system.

4. For non-equilibrium processes change of the internal energy is a function of not only thermodynamic parameters, but also of time. In the general case, equation of energy flow, the structure of which depends on the structure and state of the flow can describe change of the internal energy for open systems.

3.4. General Formulation of the First Law of Thermodynamics. Heat Capacity

1. Now the first law of thermodynamics for open systems could be formulated. However, the new formulation becomes more understandable, when an example of the interaction between the surrounding and the system with consideration for spontaneous and non-spontaneous processes proceeding in them is presented as an illustration.

Suppose that a cylinder with a plunger was chosen with the ideal gas under it (a rarefied gas under low pressure). The latter does not take into account energy consumption for intermolecular interaction during increase of the gas volume or, vice versa, release of the energy excess at gas compression.

The cylinder works under constant P and T.

Let the open system is added by heat produced during any process (burner, furnace heater, etc.).

Heat of a non-spontaneous process added to the system is consumed for heating up the working substance and the cylinder with the plunger, which causes their internal energy to be boosted $dU_{n\text{-}sp}$ (non-spontaneous process);a part of heat is released into the surrounding at the sacrifice of cooling the system (spontaneous process) Q_{sp}.

Heat released by the system into the surrounding increases the store of the internal energy of the surrounding (non-spontaneous process) and changes another parameter, which will be mentioned below. Hence totality of all spontaneous and non-spontaneous processes and their effect of the state of the system and the surrounding is reflected in the system considered.

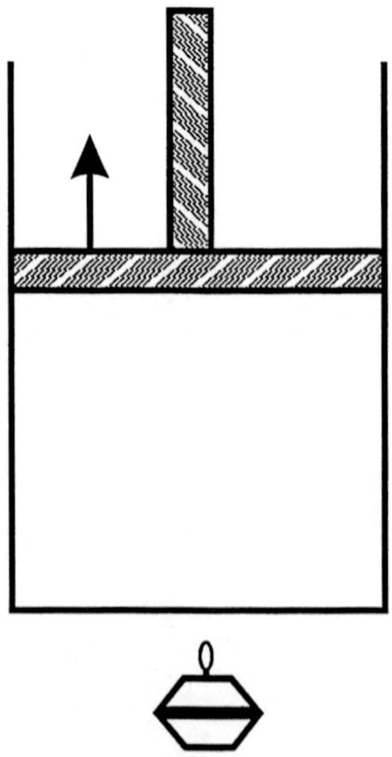

Figure 3.1. Illustration for the work performed by heated gas with the help of plunger-cylinder-working substance system and the surrounding.

2. Analytically, distribution of energy and work in the system contacting the surrounding can be generally presented as follows:

$$d_eQ_{n\text{-}sp} + d_iQ_{sp} = d_iU_{sp} + P_id_iV_{sp} + P_ed_eV + d_eU + d_eW + d_iW_{friction}. \qquad (3.10)$$

This equation represents an analytical expression for the first law of thermodynamics for open systems. Basing on this expression, the first law of thermodynamics can be formulated as follows: *Energy (heat) of a non-spontaneous process of the source, absorbed by a working substance, is consumed for an increase of the internal energy of the system (spontaneous process), conduction of work (expansion and friction), spontaneous processes and cooling down of the working substance in the system with internal energy increase in the surrounding (non-spontaneous process).*

This equation does not clearly reflect the type of processes proceeding – reversible and irreversible, but division of processes to spontaneous and non-spontaneous ones defines that all processes are thermodynamically irreversible.

3. A new parameter — time would be necessarily introduced into this equation, because all parameters of the equation are functions of it. In the mentioned view the equation is presented for equilibrium or steady states of the system. This displays its uncertainty, typical for classic thermodynamics. However, this method possesses good points, because it enables quantity of consumed or added energy and work to be determined by initial and end states of the system and the surrounding at the prescribed time.

More strictly, the equation of the first law of thermodynamics should be presented as a functional temporal dependence. However, this method is not developed yet for spontaneous and non-spontaneous conjugated processes.

4. Analytically, the first law of thermodynamics reflects a totality of simultaneous proceeding of spontaneous and non-spontaneous processes. For example, analytical expression of the first law of thermodynamics can be presented in the following form:

$$dQ_{n\text{-}sp} = dU_{n\text{-}sp} + dW_{sp}.$$

In its right part, this equation includes a parameter of the internal energy and work produced by the system. If the system absorbs energy in the form of heat (non-spontaneous process), its internal energy rises (non-spontaneous process), and the system conducts a work (spontaneous process).

Analytical expression of the first law of thermodynamics can be also presented in another form, which reflects both phenomena:

$$(\pm Q) = (\pm dU) + (\pm dW).$$

Heat is released by the working substance (spontaneous process) by virtue of reduction of a store of the internal energy (accumulated during gas compression) of the working substance (non-spontaneous process) and work addition (non-spontaneous process). Simultaneously, the left part of the equation is characterized by exchange of energy with the working substance, which adds and absorbs energy in the form of heat, i.e. it also reflects the totality of spontaneous and non-spontaneous processes (see the Appendix).

Heat Capacity

1. Heat capacity reflects a specific, local behavior of energy in the heat form in the working substance — gas, liquid or solid. In the physical sense, it defines local deviations (fluctuation) of energy value in the heat form from its average ones under current temperature. The average energy in the working substance is characterized by the value:

$$E_{av} = RT.$$

Deviations of energy from its average value in the working substance are called fluctuations. Energy fluctuations reflect a non-spontaneous process, because regularity of the fluctuation structure is higher than that of a homogeneous medium. In this case, scattering of fluctuations is a spontaneous process. The total number of such fluctuations depends on temperature T. The higher is temperature, the higher is the number of fluctuations in the particular volume of the working substance and heat capacity of the substance. Consequently, heat capacity is the value depending on temperature, and is defined as a measure of temperature change with heat amount for the mass unit of the working substance. Heat can be added to the working substance or released from it.

2. From grounds of thermodynamics of spontaneous and non-spontaneous processes, heat capacity of the working substance is determined by heat amount, added to it (non-spontaneous process) during its heating up in a spontaneous process, or released from the system into the surrounding in a non-spontaneous process.

Analytically, sign in the heat capacity equation defines type of the process:

$$\pm\left(\frac{\partial Q_v}{\partial T}\right)_v = C_v,$$

$$\pm\left(\frac{\partial Q_p}{\partial T}\right)_p = C_p,$$

$$\pm\left(\frac{\partial (U - U_0)}{\partial T}\right)_v = C_v,$$

$$\pm\left(\frac{\partial (H - H_0)}{\partial T}\right)_p = C_p,$$

where U_0 and H_0 are internal energy and enthalpy at $T = 0$ K, respectively. These values are constant and become zero after differentiation.

3.5. Enthalpy in Open Systems

1. Enthalpy represents a thermodynamic function, which includes internal energy and work of the working substance expansion. In spontaneous processes (predominance of a spontaneous process over non-spontaneous one) the value of enthalpy in the system decreases, and in non-spontaneous — rises. Enthalpy is a thermodynamic function, the value of which depends on the internal energy, thermodynamic parameters, and beside work of expansion can include other types of work. In open systems, enthalpy also depends on time of the system state change under constant pressure. For open systems, the term of thermodynamic parameter is sensible for short-time or long-time steady states of the system, because the equilibrium state cannot be reached in them. The state of an open system changes continuously by virtue of fluctuations of thermodynamic parameters and continuous exchange of mass and energy between the system and the surrounding. Nevertheless, each fixation of the state of the open system can be considered as a quasi-equilibrium state, and this enables the state and change of the system state to be calculated.

3.5.1. Mathematical Presentation of Enthalpy

1. Analytical expression of enthalpy in the above-mentioned sense can be obtained on the basis of equation of the first law of thermodynamics:

$$dQ_p = dU + PdV + dW', \tag{3.11}$$

where PdV is the work of expansion (compression), W' marks other type of work. In the absence of other types of work equation (3.11) is reduced to the type

$$dQ_p = dU + PdV. \tag{3.12}$$

Integrating this equation in the finite borders from the first to the second state, it is obtained:

$$Q_p = U_2 - U_1 + P(V_2 - V_1). \tag{3.13}$$

Grouping variables with the same indexes, the expression is obtained:

$$Q_p = (U_2 + PV_2) - (U_1 + PV_1). \tag{3.14}$$

Expression in brackets is the analytical form of enthalpy:

$$H = U + PV. \tag{3.15}$$

Substituting (3.15) into (3.14), we obtain that the difference of enthalpies equals to the process heat under p = const, $Q_p = \Delta H$. This equation determines that in open systems a spontaneous process proceeds at the sacrifice of enthalpy leakage, and by virtue of conduction of a non-spontaneous process enthalpy rises — at addition of energy to the system in the heat form, which is reflected in signs:

$$(\mp Q_p) = \mp\Delta H, \tag{3.16}$$

where negative sign determines the energy release, and positive one — energy addition from outside to the open system.

3.5.2. Potential Energy and Enthalpy

1. Generally, enthalpy can be considered as the energy of compressed or expanded working substance. In the first case, the system contains a store of potential energy, and in the second case, it is characterized by the absence of excessive store of potential energy and is determined as an expanded system.

Analytical presentation of enthalpy for the above-mentioned cases can be obtained with the help of the following transformations of thermodynamic equations. Differentiating of equation (3.15) gives

$$dH = dU + PdV + VdP, \tag{3.17}$$

where PdV and VdP determine a change of enthalpy with kinetic and potential energies of the system. Substituting (3.11) into (3.17), the following expression is obtained:

$$dH = dQ_p + VdP + \Sigma W'. \tag{3.18}$$

The sum of elementary works W' can include the following types of work: W_{diss} — the work of dissipation; gdh — the work of body lifting in the Earth's gravitation field; $-qd\varphi$ — the work of charge transfer in potential electric field; $-\varphi dq$ — the work conducted by the field under the effect of charges; σds — the work of surface tension forces; $\xi d\gamma_i$ — the work of a change of the coordination number of the working substance (structure of the working substance); HdB — the work in magnetic field; $\nu_i\mu_i d\lambda$ — the work of a chemical process; $\mu_i dn_i$ — the work of mass-exchange processes between an open system and the surrounding, and others. In the above-mentioned expressions of work the following parameters should be marked: g — the force of gravity; h — height; q — charge; φ — electric field potential; σ — surface tension; s — square of the interphase; ξ — energy of the ligand-cation bond in the polyhedron; γ — coordination number of a particle with central symmetry (ion, atom, molecule); H — magnetic field intensity; B — magnetization of the working substance; ν_i — stoichiometric coefficients; μ_i — chemical potentials of substances in the mixture; λ — chemical variable; n_i — number of moles of the substance in current phase. Enthalpy is an extensive function, and equations to calculate works are the multiplication of intensive (coefficients) and extensive parameters.

The total equation for enthalpy in a developed form will be of the following view:

$$\begin{aligned}\pm dH = \pm dQ_p \pm VdP \pm gdh \pm qd\varphi \pm \varphi dq \pm \sigma ds \pm \xi d\gamma_i \pm \\ \pm HdB \pm \Sigma_{\text{chem.react.}} \nu_i\mu_i d\lambda \pm \Sigma_{\text{ph.reconf.}} \mu_i dn_i \pm \ldots\end{aligned} \tag{3.19}$$

2. A change of enthalpy for spontaneous and non-spontaneous processes can be determined with the help of equation:

$$dQ + VdP \lesseqgtr 0. \tag{3.20}$$

Open systems are able to develop or degrade the structure in accordance with prevailing of ratios of addition or release of energy and substance to the system from the surrounding and, vice versa, from the system to the surrounding. In real systems, various physical and, in particular, mass-exchange processes proceed continuously between both phases in the system, and the system and the surrounding. These processes do also continuously determine proceeding of chemical reactions with appearance of new substances or degradation of the substances.

It should be pointed out that for intensive parameters of a chemical process and mass-exchange, chemical potentials could be used in equation (3.19), if natural variables are chosen as independent parameters. In the case, if heat in this equation is presented by the following expression:

$$dQ_p = C_p dT, \tag{3.21}$$

partial molar values are introduced into equation (3.19) instead of chemical potentials:

$$\overline{H} = \left(\frac{\partial H}{\partial n_i}\right)_{p,T,n_{j\neq i}} . \tag{3.22}$$

Equation (3.19) will be written down in the following form:

$$\begin{aligned} \pm dH &= \pm dQ_p \pm VdP \pm gdh \pm qd\varphi \pm \varphi dq \pm \sigma ds \pm \xi d\gamma \pm \\ &\pm HdB \pm \sum \overline{H}_r dn_r \pm \sum \overline{H}_m dn_m \pm \dots \end{aligned} \tag{3.23}$$

where r is the reaction, m is the mass-exchange.

For open systems the following indexes are introduced as before: i — internal and e — external parameters. Open systems can be subdivided into inorganic, organic and biological ones. The grade of these systems grows in the mentioned sequence. In nature, inorganic systems are the most widespread, followed by organic ones, and the sequence ends by biological systems (of vegetable and animal origin). In these systems exchange processes between internal parts of systems and internal and external parts of them proceed continuously, with these processes including intimate combination of spontaneous and non-spontaneous processes. The systems are also simultaneously affected by a great number of different in origin i- and e-parameters. To simplify the study of behavior of the systems, the effects of some parameters on the system can be neglected (conditionally), and the equation for enthalpies can be presented in a more simple form:

$$dH = \pm dQ_p \pm VdP \pm \sum \overline{H}_r dn_r \pm \sum \overline{H}_m dn_m . \tag{3.24}$$

3. For open systems variables in equation (3.24) are separated for internal and external processes:

$$d_iH + d_eH = d_iQ + d_eQ + V_id_iP + V_ed_eP + \sum \overline{H}_i d_i n_r + \\ + \sum \overline{H}_i d_e n_r + \sum \overline{H}_i d_i n_m + \sum \overline{H}_i d_e n_m \quad (3.25)$$

In this equation, the term $\sum \overline{H}_i d_i n_r$ determines the work of chemical processes proceeding inside the system; $\sum \overline{H}_i d_e n_r$ — the same in the surrounding; $\sum \overline{H}_i d_i n_m$ determines the mass-exchange between phases inside the system; $\sum \overline{H}_i d_e n_m$ determines the mass-exchange between the system and the surrounding.

Let us separate variable for internal medium and the surrounding:

$$d_iH + d_eH = \left(d_iQ + V_id_iP + \sum \overline{H}_i d_i n_r + \sum \overline{H}_i d_i n_m\right) + \\ + \left(d_eQ + V_ed_eP + \sum \overline{H}_i d_e n_m + \sum \overline{H}_i d_e n_r\right) \quad (3.26)$$

To determine the type of the process, this equation can be integrated in the range from the first to the second and from the second to the first state. The first term determines physicochemical processes, proceeding inside the system and interacting with the surrounding. The second term determines processes proceeding in the surrounding and interacting with the open system.

3.6. Change of Enthalpy and Thermochemistry as a Section of Thermodynamics. Designations in Thermochemistry

Thermochemistry as a section of chemical thermodynamics studies regularities of proceeding of chemical reactions and accompanied heat effects on the basis of G.I. Gess law. This law reflects the main property of thermodynamic functions – independence of change of the internal energy and enthalpy on the way of the process proceeding, unambiguity and continuity of the function change.

Chemical reactions can be subdivided into spontaneously and non-spontaneously proceeding ones. Usually, spontaneously proceeding reactions are accompanied by energy release in the heat form. These reactions are called exothermic, and the negative sign is assigned to enthalpy. However, some reactions proceed spontaneously, but absorbing energy in the heat form. If considered in detail, spontaneously proceeding reactions can be also separated in these reactions. Among reactions of the first type are those of synthesis of organic and inorganic compounds from simple substances. The following reactions with the exothermal effect can be displayed:

$Fe + O_2 = FeO_2 - 95400$ J/mol,

$H_2 + Cl_2 = 2HCl - 92300$ J/mol.

However, reactions of the type

$H_2 + I_2 = 2HI + 25940$ J/mol,
$NO_2 + NO_2 = N_2O_4 + 9730$ J/mol

are endothermic reactions. The main obstacle of proceeding of these reactions is an energetic barrier. Reactions become spontaneous after overcoming it.

Usually, non-spontaneous processes proceed with absorption of energy in the heat form:

$2C + 2H_2 = C_2H_4 + 52283$ J/mol,
$6C + 3H_2 = C_6H_6 + 82927$ J/mol.

Such reactions as synthesis of butenes, cyclohexanes, cyclopentanes, *n*-heptane cannot proceed spontaneously, although heat effect of the reactions is negative, i.e. it is the exoreaction. The main role in these processes is also played by stages of activation of molecules of initial substances, after which reactions proceed spontaneously.

It will be shown below that the type of proceeding chemical process – spontaneous or non-spontaneous – can be strictly determined by the sign of change of Gibbs or Helmholtz energy.

Designations in the Thermochemistry

1. The following designations are introduced in the thermochemistry for the sake of convenience. Simple substances are called the simplest compounds of elements, existing in a stable form under common (Earth) conditions: H_2, O_2, C_{solid}, S, N_2, and others. Complex compounds include two or more different elements: HCl, H_2SO_4, C_2H_4, C_2H_5Cl, etc.

2. The heat of formation of compounds from simple substances is determined by the energy amount, released or added during the synthesis of the compound from simple substances:

$H_2 + 1/2O_2 = H_2O - 241840$ J/mol.

The heat effect of the chemical reaction is determined by energy in the heat form, released or added during chemical interaction of complex compounds or complex and simple substances with each other:

$C_2H_4 + H_2O = C_2H_5OH - 45708$ J/mol,
$C_2H_4 + H_2 = C_2H_6 - 32384$ J/mol.

Atomic heat is determined by the amount of energy, released during formation of compounds from atoms:

$2H = H_2 - 434299$ J/mol,

$2O = O_2 - 491201$ J/mol,
$2H + O = H_2O - 921735$ J/mol.

Atomic heat of formation of a compound from atoms can be considered as the energy of bonds of atoms in a molecule.

Combustion heat is determined as an amount of energy, released during complete combustion of 1 mole of a substance up to formation of higher oxides. Lower and higher combustion heats of substances are differed. The higher combustion heat is determined by the amount of heat, released during complete combustion of one gram-mole of a substance under complete condensation of water vapors formed during combustion of a hydrogen-containing fuel. The lower combustion heat determines the heat amount, released during complete combustion of substances up to formation of gaseous products with no condensation of water vapors.

Heats of dilution are subdivided into integral and differential ones. Integral heat of dilution is the amount of heat, released or absorbed in the solution, when n_1 moles of dissolved substance are mixed with n_2 moles of solvent under constant P and T. This is the heat of mixing of pure substances with each other. Differential heat of dissolution is determined as the amount of heat, released or absorbed by a solution during dissolving of an infinitely small amount of a substance in it under constant P, T and n_i. This value is called the partial molar heat of dissolution.

The dissolution heat of the present substance in infinitely diluted solution is called the first heat of dissolution, and in saturated solution – the final heat of dissolution.

3. Under constant volume heat is called the heat at V = const, and under constant pressure – the heat at P = const:

$$\Delta H = Q_v, \tag{3.27}$$

$$\Delta H = Q_p. \tag{3.28}$$

For the sake of convenience of calculations in thermochemistry, standard values of heats are used, which respond to the pressure of 1 atmosphere, 101.3 kPa, 1.0133×10^5 Pa; temperature is arbitrarily equaled to 298.15 K. In this case, the substance can exist in any aggregate state – gaseous, liquid or solid. In the most of cases the substance state is hypothetical under standard pressure and is determined with the help of methods created in thermodynamics.

3.7. THE LAW BY G.I. GESS

1. The law by G.I. Gess characterizes constancy of sums of heats for reactions proceeding through various stages, but between the same initial substances and reaction products. Until the present time no attention was paid to a possibility to separate spontaneous and non-spontaneous chemical reactions among the total complexity of the reactions. Spontaneous reactions are characterized by subsidence of the free energy and rise of entropy, and non-spontaneous – by rise of a store of the free energy and subsidence of entropy. Moreover, spontaneous reactions are usually accompanied by release of energy in the heat form, i.e. the process is exothermic.

However, chemical reactions can be also noted, which proceed spontaneously, but with absorption of energy in the heat form (for example, the reaction of the Glauber's salt with hydrochloric acid).

Initial substances in exothermal reactions possess higher store of energy, than reaction products, as, for example, in the processes:

$H_2 + 1/2O_2 = H_2O - 241840$ J/mol,
$H_2 + Cl_2 = 2HCl - 92300$ J/mol,
$Fe + 1/2O_2 = FeO - 263680$ J/mol,
$C_2H_4 + H_2O = C_2H_5OH - 45708$ J/mol.

Interaction of such a system with the surroundings causes transfer of the energy excess in the heat form to the surroundings and its heating.

The reverse reactions are non-spontaneous and can proceed only under energy addition from the outside (heating, radiation), enough to transmit energy to initial substances in the amount exceeding their energy (for example, for the reaction type of $H_2O = H_2 + 1/2O_2$). The example of endothermic reaction can be isobutylene isomerization into *n*-butylene:

iso-C_4H_8 = *n*-C_4H_8 + 3900 J/mol.

This reaction can proceed under increased temperatures only, when added heat exceeds the reaction one. Isomerization of isobutylene is considered as a non-spontaneous process. At $T = 298$ K the energy of *n*-butylene is higher than that of isobutylene, and the reaction does not proceed directly at this temperature.

2. Most often, chemical reactions are multistage ones. In this case, a totality of spontaneous and non-spontaneous processes between initial substances and reaction products of the same composition can proceed by different directions.

Spontaneous processes induce proceeding of chemical reactions, and non-spontaneous ones decelerate them. These two sides of processes are interconnected and continuous. These processes are called conjugated. Simple example of conjugated chemical reactions can be dehydrogenation of paraffin hydrocarbons in the presence of O_2:

$n\text{-}C_4H_{10} = n\text{-}C_4H_8 + H_2 + 125870$ J/mol,
$H_2 + 1/2O_2 = H_2O - 241840$ J/mol.

In this combination of reactions, the first one proceeds with absorption of energy in the heat form (endothermic reaction), and the second one – with release of energy in the heat form (exothermic reaction). In this case, heat of the hydrogen oxidation is higher by its absolute value than that of dehydrogenation of *n*-butane. In this conjugated pair of reactions, oxidation of H_2 is quite enough for butane dehydrogenation.

Hence the law by G.I. Gess can be formulated in a more concrete form: the heat effect of conjugated spontaneous and non-spontaneous processes does not depend on the way of the process proceeding (on the number of intermediate stages), but depends on the initial and the end states of the systems only under P and T = const or V and T = const.

3. Analytically, the law by G.I. Gess can be presented for several ways of proceeding of the reactions between the same initial substances and reaction products as follows:

$$\Delta H_p = \Sigma\nu_i\Delta H_i = \Sigma\nu_j\Delta H_j = \Sigma\nu_k\Delta H_k = \pm(\Delta H_p), \tag{3.29}$$

where indexes i, j, k reflect ways of proceeding of chemical reactions, different by nature from the intermediate stages. To keep temperature constant in the system during an endothermic process, addition of energy in the heat form from the surrounding is necessary. For an exothermic process, the excess of heat is released from the system into the surrounding, which is reflected by the sign in equation (3.29) in expressions of the following type:

$$(-\Delta H_{chem.react.}) = \Delta H_{surround.},\ \Delta H_{chem.react.} = -\Delta H_{surround.}.$$

Heats of chemical reactions are determined for those aggregate states, in which the substance exists under normal conditions. For the sake of convenience to calculate heats of the processes of different types (should be outlines once more) it was agreed to take pressure, under which the heat effect is measured (usually, it is calculated for standard conditions), equal to 1 atmosphere (101.3 kPa), and temperature T = 298 K (temperature is not standardized), finally presented in the form: ΔH^0_{298}.

Combustion heats for substances and formation heats for those, for which chemical reaction can be conducted by a simple method, are experimentally determined with the help of calorimetric bombs, differential scanning calorimeter and other methods.

At the present time, types of reactions — spontaneous and non-spontaneous — are neglected in calculations of heats of chemical reactions (heats of combustion and formation, heat effects). Heats of chemical reactions are additive parameters, and this is bound to the following equations:

$$Q_v = \pm\Delta U, \tag{3.30}$$

$$Q_p = \pm\Delta H. \tag{3.31}$$

For these particular cases heat is the function of state and, as a consequence that U and H are additive functions, heats are also additive. Chemical reactions in combination with heat effects are called thermochemical equations. These equations can be summarized and subtracted, multiplied and divided by stoichiometric parameters.

4. The following rules of calculation of formation heats and heat effects of chemical reactions were formulated on the basis the additive property of thermodynamic functions.

Heat of formation of compounds from simple substances can be calculated by subtracting totality of combustion heats of reaction products from the sum of combustion heats of initial substances taking into account stoichiometric effects. Analytically, this is presented by the equation

$$\left(\Delta H_{298}^{0}\right)^{form} = \sum v_i \left(\Delta H_{298}^{comb,0}\right)_{init.subs.} - \sum v_i' \left(\Delta H_{298}^{0,comb}\right)_{products}, \qquad (3.32)$$

where v_i, v_i', $\sum v_i \left(\Delta H_{298}^{comb,0}\right)_{init.subs.}$, $\sum v_i' \left(\Delta H_{298}^{0,comb}\right)_{products}$ are stoichiometric coefficients and heats (enthalpies) of combustion of initial substances and reaction products; $\left(\Delta H_{298}^{0}\right)^{form}$ is the heat of formation of a compound from simple substances. The sign ^{0} (zero) determines the standard state for the substance, i.e. pressure of 1 atmosphere.

It should be noted that equations (3.30) and (3.31) display the connection between the heat effect at V = const and P = const and changes of the internal energy and enthalpy. Right parts of equations reflect processes proceeding in the system, and left parts characterize heats transferred from the system into the surrounding and back.

5. The second rule characterizes the method of calculation of heat effects of chemical reactions according to heats of formation. In accordance with this rule, the heat effect of a chemical reaction equals to the sum of heats of reaction products' formation minus the sum of formation heats of initial substances with consideration for stoichiometric coefficient under constant V and T or P and T:

$$\left(\Delta H_{298}^{0}\right)_{react} = \sum v_i' \left(\Delta H_{298}^{0}\right)_{products}^{form} - \sum v_i \left(\Delta H_{298}^{0}\right)_{init.subs}^{form}. \qquad (3.33)$$

6. The third rule concerns the atomic heat of formation: the formation heat is calculated by difference of sums of connection energies of initial substances and reaction products and heats of the phase transformation of substances:

$$\left(\Delta H_{298}^{0}\right)_{atomic}^{form} = \sum v_i' \left(\Delta H_{298}^{0}\right)_{init.bond}^{form} + \Delta H_{ph.transf.} - \sum v_i \left(\Delta H_{298}^{0}\right)_{prod.bond}^{form}, \qquad (3.34)$$

where $\left(\Delta H_{298}^{0}\right)_{atomic}^{form}$ is the atomic heat of the compound formation from combination of appropriate atoms; $\left(\Delta H_{298}^{0}\right)_{init.bond}^{form}$, $\left(\Delta H_{298}^{0}\right)_{prod.bond}^{form}$ are average bond energies in initial substances and reaction products.

As follows from the above-discussed rules, heats of reactions are calculated for temperature $T = 298$ K only. To recalculate heats from this temperature to any other, the Kirchhoff formula is used.

3.8. Temperature Dependency of the Heat Effect

1. In nature and technology, chemical reactions usually proceed at temperatures different from $T = 298$ K. Heat effects under different temperatures can be calculated by the Kirchhoff formula, which can be easily deduced by the method of thermodynamic functions. In the general case, chemical reaction can be presented by a stoichiometric equation as follows:

$$\sum v_i A_i = \sum v_i' A_i', \tag{3.35}$$

where v_i, v_i', A_i, A_i' are stoichiometric coefficients and symbols of initial substances and products of the chemical reaction. Initial substances and reaction products are characterized by energies H_{A_i} and $H_{A_i'}$, respectively, and change of energy (a store of the potential energy) during chemical reaction is determined from the difference

$$\Delta H_p = \sum v_i' H_{A_i'} - \sum v_i H_{A_i}, \tag{3.36}$$

where $H_{A_i}, H_{A_i'}$ are enthalpies of initial substances and reaction products.

Differentiating equation (3.36) by temperature under P = const, we'll obtain the following expression:

$$\left(\frac{\partial \Delta H_p}{\partial T}\right)_p = \sum v_i' \left(\frac{\partial H_{A_i'}}{\partial T}\right)_p - \sum v_i \left(\frac{\partial H_{A_i}}{\partial T}\right)_p. \tag{3.37}$$

Partial derivatives of enthalpy by temperature under constant pressure equal to the heat capacity at P = const for initial substances and reaction products:

$$\left(\frac{\partial H_{A_i}}{\partial T}\right)_p = Cp_{A_i}, \tag{3.38}$$

$$\left(\frac{\partial H_{A_i'}}{\partial T}\right)_p = Cp_{A_i'}. \tag{3.39}$$

2. As mentioned in the previous Section, exothermic reactions proceeding in open and closed systems release the energy excess into the surrounding or heat up the system under insufficient heat exchange. Evidently, if temperature in the open system T_{sys} is higher than of the surrounding $T_{surround}$, a spontaneous energy transfer in the heat form from the system into the surrounding happens, which is characterized by the following inequality:

$$\left(\frac{\partial \Delta H_{sp}}{\partial \tau}\right)_p > \left(\frac{\partial \Delta H_{n-sp}}{\partial \tau}\right)_p \tag{3.40}$$

or

$$\left(\frac{\partial \Delta H_{sp}}{\partial \tau}\right)_p - \left(\frac{\partial \Delta H_{n-sp}}{\partial \tau}\right)_p = \upsilon_{cool}^{H}. \tag{3.41}$$

Cooling of an open or closed system proceeds up to system temperature leveling with the surrounding.

Endothermic reactions in closed systems cool down the working substance. Temperature of the working substance $T_{w.s.}$ will be lower than that of the surrounding $T_{surround}$, and the chemical system will attract energy in the heat form from the surrounding. Non-stationary conditions are characterized by the following inequalities:

$$\left(\frac{\partial \Delta H_{sp}}{\partial \tau}\right)_p < \left(\frac{\partial \Delta H_{n-sp}}{\partial \tau}\right)_p \tag{3.42}$$

or

$$\left(\frac{\partial \Delta H_{sp}}{\partial \tau}\right)_p - \left(\frac{\partial \Delta H_{n-sp}}{\partial \tau}\right)_p = \upsilon_{heating}^{H}. \tag{3.43}$$

Consequently, the system will be heated up by energy added from the surrounding to the chemical system at the rate $\upsilon_{heating}$. These correlations will be taken into account as signs in the left and right parts of the Kirchhoff equation, which can be deduced already by substituting equations (3.38) and (3.39) into (3.37). After that the temperature dependence will be the following:

$$\left(\frac{\partial \Delta H_T}{\partial T}\right)_p = \sum v_i' C_{P_{A_i'}} - \sum v_i C_{P_{A_i}} . \tag{3.44}$$

In more simple view, this equation is obtained in the form:

$$\left(\frac{\partial \Delta H_T}{\partial T}\right)_p = \pm \Delta C_P , \tag{3.45}$$

where

$$\Delta C_P = \sum v_i' C_{P_{A_i'}} - \sum v_i C_{P_{A_i}} , \tag{3.46}$$

and $C_{P_{A_i'}}, C_{P_{A_i}}$ are heat capacities of initial substances and reaction products, J/mol·K.

Equation (3.44) is used to calculate heat of a chemical process under different temperatures. Variables are separated in equation (3.45) or (3.44), and the expression obtained is integrated in the range from 0 to T or from T_1 to T_2:

$$\Delta H_T^0 = \Delta H_{T=0}^0 + \int_0^T \Delta C_P dT , \tag{3.47}$$

$$\Delta H_{T_2}^0 = \Delta H_{T_1}^0 + \int_{T_1}^{T_2} \Delta C_P dT . \tag{3.48}$$

Equations of heat capacity dependence on temperature in the exponential form are substituted into the previous ones:

$$C_P = a + bT + cT^2 , \tag{3.49}$$

or in the theoretical form (a, b, c are empirical coefficients). Then analytical or numerical computerized integrating is performed.

3. Exothermic chemical processes in a closed system, with insufficient heat exchange with the surrounding, rise temperatures of the working substance. The rate of temperature rise is determined as the first temporal derivative of heat capacity or the second derivative of enthalpy, as shown below:

$$-\left(\frac{\partial \Delta H}{\partial T}\right)_p = \Delta C_P, \tag{3.50}$$

$$-\left(\frac{\partial \Delta C_P^{sp}}{\partial \tau}\right)_p = \upsilon_{T,sp}, \tag{3.51}$$

$$-\left(\frac{\partial^2 H_{sp}}{\partial \tau \partial T}\right)_p = \upsilon_{T,sp}. \tag{3.52}$$

In this case, the store of energy in molecules of initial substances is higher than that in molecules of reaction products and the energy excess in the heat form is transferred to the surrounding. As this takes place, simultaneous proceeding of spontaneous processes (cooling) and heating up of the surrounding (non-spontaneous process) determine proceeding of a chemical reaction at a finite rate. Inequalities of the following type can be suggested for this case:

$$\Delta H_{sp}^{sys} > \Delta H_{n-sp}^{surround}, \tag{3.53}$$

$$\left(\frac{\partial \Delta H_{sp}^{sys}}{\partial \tau}\right)_p > \left(\frac{\partial \Delta H_{n-sp}^{surround}}{\partial T}\right)_p, \tag{3.54}$$

$$\left(\frac{\partial^2 \Delta H_P^{sp}}{\partial \tau \partial T}\right)_p > \left(\frac{\partial^2 \Delta H_P^{n-sp}}{\partial \tau \partial T}\right)_p, \tag{3.55}$$

$$\left(\frac{\partial \Delta C_{P,sys}^{sp}}{\partial \tau}\right)_p > \left(\frac{\partial \Delta C_{P,surround}^{n-sp}}{\partial \tau}\right)_p. \tag{3.56}$$

These inequalities are hardly understandable yet, but they are fulfilled phenomenologically and experimentally in all these cases.

4. For endothermic reactions, a lack of energy in a closed or open system is filled up by addition from the surrounding. This condition is determined by the following inequalities:

$$\Delta H_{sys}^{sp} < \Delta H_{surround}^{n-sp}, \tag{3.57}$$

$$\left(\frac{\partial \Delta H_{sys}^{sp}}{\partial T}\right)_p < \left(\frac{\partial \Delta H_{surround}^{n-sp}}{\partial T}\right)_p, \tag{3.58}$$

$$\left(\frac{\partial \Delta C_{P,sys}^{sp}}{\partial \tau}\right)_p < \left(\frac{\partial \Delta C_{P,surround}^{n-sp}}{\partial \tau}\right)_p. \tag{3.59}$$

These processes are thermodynamically irreversible and strive to bring the system into the equilibrium state with the surrounding. However, different processes proceed simultaneously in the surrounding, which continuously deflect the system from the state of equilibrium.

Chapter 4

THE SECOND LAW OF THERMODYNAMICS AND MAIN MATHEMATICAL EQUATIONS

4.1. INSUFFICIENCY OF THE FIRST LAW OF THERMODYNAMICS TO CLASSIFY PROCESSES

1. The first law of thermodynamics, grounded on the law of conservation of energy and quantity of motion, binds together heat and work with due regard for the law of equivalence. It enables a store of energy in the system to be determined on the basis of two thermodynamic functions — the internal energy U and enthalpy H. Analytically, to use P, V and T as variables at constant mass of the working substance, these correlations can be presented in the following view:

$$\Delta U = Q_v \quad (4.1), \qquad \Delta H = Q_p \quad (4.2),$$

$$W_v = Q_v \quad (4.3), \qquad W_p = Q_p \quad (4.4),$$

$$dW_v = PdV \quad (4.5), \qquad dW_p = VdP \quad (4.6).$$

In the general case, under participation of other intensive X_i and extensive x_i parameters in conduction of work, the latter can be expressed via the product of these parameters:

$$dW_k = \Sigma X_i dx_i. \quad (4.7)$$

In these equations plus and minus signs can be introduced before work and heat. The *minus* sign will determine heat evolution from the system into the surrounding and conduction of work by the system with creation of changes in the surrounding. The *plus* sign will determine heat transmission from the surrounding to the system and consumption of the work, which makes changes in the present system. The signs can be exchanged, but this will just change explanation of the processes, and will not change the entity of systems interacting with the surrounding and processes proceeding in them.

Hence, in the general case, type of the process and direction of its proceeding cannot be determined by signs of work and heat. Even if the signs determining consumption and evolution of heat and work are strictly agreed, obstacles in determination of direction of the process proceeding occur.

2. Basing on experimental observations, it was stated that work is always finished by its transformation into heat, with strictly equivalent relations. At the same time, heat is not always the source of work. For example, if a gas is heated up in a cylinder under a plunger, it will conduct a work, lifting the plunger in the cylinder. For the ideal gas, this work under isobaric conditions will be proportional to temperature:

$$W_p = R(T_2 - T_1). \tag{4.8}$$

But if a stone of mass m is heated up on the Earth's surface with the help of a blowtorch, for example, it will not lift itself up above the surface, if the heat is not preliminarily transformed into mechanical work. Consequently, heat can be the source of work only under specific conditions of its effect of the working substance. If an isolated object is simply heated, this just increases its internal energy, but not transforms heat into work. The process of stone lifting is the non-spontaneous one, and stone falling down and conduction of work at its contact with the Earth's surface is the spontaneous process. These two types of the process cannot be differed with the help of the first law of thermodynamics.

To separate types of processes — spontaneous and non-spontaneous ones, additional conditions are necessary, which are formulated with the help of the second law of thermodynamics and thermodynamic functions, obtained on the basis of these two laws.

4.2. Spontaneous and Non-Spontaneous Processes

1. In nature and technology, spontaneous and non-spontaneous processes can be easily separated. The former ones proceed at the sacrifice of existence of excess of the free internal energy in the working substance. These processes can be the source of work, appearance of other forms of energy (electric, chemical, biological, etc.) or the source of proceeding of a thermal process.

Proceeding of non-spontaneous processes requires consumption of work, energy in the heat form (heating up) or in other forms (electric, potential, radiation, etc.).

As an example, processes proceeding spontaneously in nature and technology are the following: heat transfer from hot working substance (from heater) to cold one (cooler), diffusion of substances from concentrated solutions to diluted ones, pressure reduction from higher to lower one, falling of solids in the Earth's gravitation field, display of osmotic pressure, and many other processes.

Processes, reverse to the above-mentioned ones, can be referred as non-spontaneous: cooling of the working substance below temperature of the surrounding, heating up of the working substance from $T_{surround.}$ to $T > T_{surround.}$, compression of gases, steaming of liquids, concentration of solvents, lifting of weights in the Earth's gravitation field, solvent elimination from the solution, etc.

These two types of processes (spontaneous and non-spontaneous) are continuous and conjugated. They proceed at a definite rate both directly (spontaneous) and reversibly (non-spontaneous). This is determined by the following difference in rates:

$$\upsilon_{total} = \upsilon_{direct} - \upsilon_{reverse}, \tag{4.9}$$

where $\upsilon_{total}, \upsilon_{direct}, \upsilon_{reverse}$ are rates of a process (spontaneous or non-spontaneous) prevailing in the system, direct process (condensation, heat transmission and other) and reverse (cooling, compression, etc.).

Usually, in nature and technology υ_{total} is of finite value. That is why such processes are thermodynamically irreversible. If the total rate of the process is infinitely low, i.e.

$$\upsilon_{total} \cong 0, \tag{4.10}$$

such processes proceed during infinitely long time, are infinitely slow, and are named quasi-equilibrium or quasi-static ones. In the classic thermodynamics, in the border case, these processes are called thermodynamically reversible or equilibrium. Following from equation (4.10), the state of equilibrium can be considered as the border or ideal change of the process state. In thermodynamics of equilibrium processes it is assumed that transition of all processes from one state into another is equilibrium. Thus, in the thermodynamics of equilibrium processes they cannot be subdivided into spontaneous and non-spontaneous ones. In equilibrium systems processes are conducted with continuous compensation of various changes in the state of the working substance, and direction of the process proceeding under such conditions can be hardly determined. However, this idea is a suitable scientific abstraction of thermodynamic study of the state of working substances and changes in it. The rate of proceeding of thermodynamic processes is studied in thermodynamics of irreversible processes. However, signs in thermodynamic equations reflect the type of the process.

2. The second law of thermodynamics applied to equilibrium processes is formulated as the principle of existence of entropy, which can be strictly defined for open systems, in which compensation is possible.

In essence of formulations of the second law of thermodynamics, which will be discussed below, the thermodynamics study irreversibly proceeding processes, which is reflected in methods and content of thermodynamics of irreversible processes. Thermodynamics of irreversible processes formulate the principle of entropy growth, which characterizes the system approaching the state of equilibrium at a finite rate, with this principle characterizing only spontaneous processes in the absence of partial or full compensation.

However, these two principles do not fully describe the processes proceeding in nature. Beside spontaneous processes, non-spontaneous ones proceed continuously in nature and technology and are conjugated to the former. These processes limit the rate of their combined proceeding and represent their integral property. That is why a principle of entropy decrease and rise of the free energy in non-spontaneously proceeding processes should add both these principles.

3. In nature, phase-open, adiabatic and closed systems can be separated. The former ones exchange energy and substance with the surrounding, the second — substance only, and the latter — energy only. Hence interaction of spontaneous and non-spontaneous processes in these systems are characterized by equal level of the store of energy in the system and in the surrounding, and by their different nature. In the phase-open systems, changes on energetic and material levels proceed simultaneously, which reflects the same, but in an opposite direction, changes in the surrounding. For example, if a system possesses an excess of energy and substance comparing with the surrounding, it spontaneously evaluates this excess into the surrounding, and the latter decelerates these processes by those opposite to them, i.e. by the opposite influence of the material and energetic composition of the surrounding.

4. Spontaneous processes proceed at the sacrifice of reduction of energy and substance excesses (chemical potential) in the system and lead to degradation of the system energy. In their turn, non-spontaneous processes proceed at the expense of work or energy only for changing structure and state of the system. Accumulation of energy and substance in the system determines rise of potential abilities of the systems, and vice versa. That is why degradation and development of the system proceed continuously and are conjugated as a result of proceeding of spontaneous and non-spontaneous processes.

Clearing up the main types of conjugated spontaneous and non-spontaneous processes, their nature, conditions of their proceeding, working substances participating in interaction between the system and the surrounding, the main regularities of their interaction can be stated, processes can be mathematically described, preferable direction of the resulting process proceeding, characterized by the rate υ_{total} can be cleared up. New approach will enable problems of degradation degree and development of systems and the surrounding to be solved, as well as problems of evolution of systems, the level of evolution, etc. These directions in thermodynamics of irreversible processes require the axiom solution, creation or modification of the mathematical apparatus and solution of problems bound to interaction of spontaneous and non-spontaneous processes in various systems, including isolated ones, from new positions.

4.3. ILLUSTRATION OF A PAIR OF CONJUGATED PROCESSES IN THE FRAMES OF A SINGLE SYSTEM

Proceeding and interaction of a pair of processes — spontaneous and non-spontaneous — can be most simply illustrated on the example of the Gay-Lussacs—Joule experiment, described in Section 2.2.3. In this experiment, two sphere-like volumes are connected by a pipe with a plug valve. They possess a branch pipe to fill in the volume by gas or to pump it out down to low residual pressure, as shown above in Figure 2.2.

Vessel 1 is filled by hydrogen up to the pressure of 10-15 atmospheres, and air was pumped out from vessel 2 down to residual pressure of 0.1 atmosphere (0.01 MPa). Both vessels were placed into a thermostat. If valve 9 is open, hydrogen will overflow from the first vessel into the second one. In this case, pressure will decrease in the first vessel and boost in the second one. The process of hydrogen overflowing from the first vessel to the second will be spontaneous. Pressure in vessel 1 will decrease from P_1 to P_{eq}, and temperature — from T_1 to T^*_{eq}. In the second vessel, a non-spontaneous process of pressure boosting from P_1' to P_{eq} and temperature growth from T_1 to T_{eq} takes place. In the first case, $P_1 > P_{eq}$, $T_1 < T^*_{eq}$, and in the second vessel $P_1' < P_{eq}$, $T_1 > T_{eq}$. Change of entropy in the first vessel rises at the sacrifice of pressure decrease (spontaneous process) and decreases at the sacrifice of temperature reduction (non-spontaneous process) below T_{eq} (temperature of the thermostat), which can be determined from equations:

$$\Delta S_1 = -R\ln P_1/P_{eq} > 0,\ \Delta S_1' = C_p\ln T_1/T_{eq} < 0. \quad (4.11)$$

In the second vessel, entropy decreases at the sacrifice of pressure boosting and grows at the sacrifice of temperature increase, which is determined by expressions:

$$\Delta S_2 = -R\ln P_2/P_{eq} < 0,\ \Delta S_2' = C_p\ln T_2/T_{eq} > 0. \quad (4.12)$$

Changing of entropy in the medium surrounding vessel 1 will also decrease at the sacrifice of temperature reduction due to release of energy in the heat form to the sphere 1. Changing of entropy around sphere 2 will increase due to heating of the medium.

Consequently, it is clearly observed from these correlations that spontaneous process is compensated by a non-spontaneous one, and non-spontaneous – by a spontaneous one. This reflects the process of redistribution of entropy and energy in the system itself and at its interaction with the surrounding. Inequalities (4.11) – (4.12) can be used in thermodynamics of irreversible processes only.

4.4. Reversible and Irreversible Processes

1. Equiponderantly proceeding processes are called thermodynamically reversible ones, and non-equiponderantly proceeding ones – thermodynamically irreversible. These two processes differ by conditions of conduction (compensation) and rate of proceeding (finite value for irreversible processes). Under real conditions, processes proceed at a finite rate in the surrounding (and in technology) and, consequently, are thermodynamically irreversible. Usually, spontaneous processes are recognized as proceeding directly, and non-spontaneous ones – in the reverse direction, but such division of processes is not strict. The opposite division of processes can be also assumed.

Thermodynamically irreversible processes can be reduced to thermodynamically reversible ones under specific conditions of their conduction, i.e. under the condition of complete compensation of a spontaneous process by a non-spontaneous one on any elementary stages of change of the state of the working substance and the surrounding. Therewith, the necessary condition is the infinitely low rate of the process proceeding, provided by infinitely low effects of external and internal parameters on the working substance. Under this effect on the system, it is impossible to determine the type of prevailing process – spontaneous or non-spontaneous one. That is why conduction of the process under these conditions at infinitely short removal from equilibrium, their division into spontaneous and non-spontaneous ones is not possible. Processes are subdivided into spontaneous and non-spontaneous only for finite rates of proceeding.

2. Conditions of irreversible and reversible conduction of processes can be illustrated on the example of compression and expansion of the ideal gas in a cylinder with a plunger having no weight with no friction with the cylinder. Sequence of conduction of compression and expansion of the gas in the ideal state is shown in Figure 4.1 for isothermal conditions in coordinates $P_T = P_T(\upsilon)$. Irreversible compression and expansion of gas is displayed in Figure 4.1 by curves I and I′, and thermodynamically reversible – by curve II. Initially, compression of gas in the cylinder with the help of moving plunger without friction can be performed by loads of big weight. Curve I in the Figure depicts gas compression by loads of big weight.

Gas pressure boosts sharply, when loads are placed upon the plunger, and then its volume decreases abruptly. In point 2 gas transits into the equilibrium state at T = const. Broken lines in the Figure are conditionally depicted by parts of straight lines, although they can be presented differently. This is determined by the fact that non-controlled disturbances appear in gas under a finite rate of its compression: friction between gas layers, turbulization /agitation/, fluctuation of density, etc. Similarly, the process is conducted through points 3 and 4.

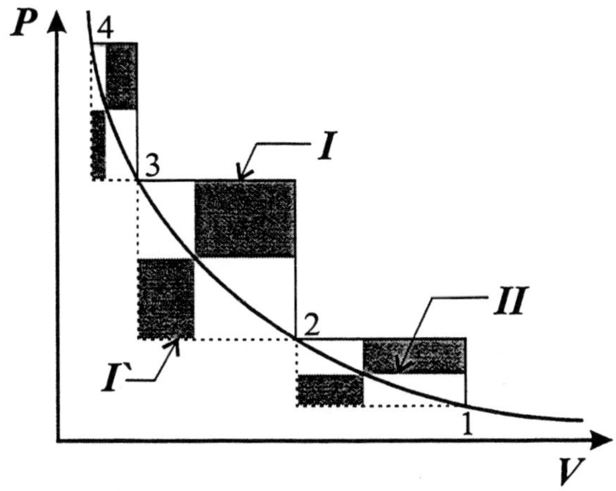

Figure 4.1. Illustration of compression and expansion of the ideal gas in irreversible (I, I′) and reversible (II) processes.

Square under the broken curve I – compression of gas – reflects work spent for gas compression $W_{compr.}$. Then the process is conducted irreversibly, in the opposite direction. For this purpose, first load is taken off from the plunger surface. Gas pressure decreases abruptly under the plunger, and then its volume increases also sharply from point 4 to point 3. The process is conducted in the reverse direction up to the initial point 1. It can be noted that in the reverse direction gas expansion also proceeds through equilibrium points 4, 3, 2, and 1. The square below the broken curve I′ - gas expansion – determines the work conducted by gas at its expansion, $W_{expans.}$. These two processes were conducted irreversibly, and works are different:

$$W_{expans.} < W_{compr.}. \tag{4.13}$$

The square under curve I is greater than that under curve I′, and work spent for gas compression is greater than that conducted by gas at expansion.

Non-coincidence of curves I and I′ of irreversible processes of gas compression and expansion happens due to incomplete compensation of a non-spontaneous process by a spontaneous one, and vice versa. This is the very fact that determines irreversibility of the processes proceeding at a finite rate.

Decreasing the size of loads, the number of points can be increased, which reflects equilibrium states of the system at compression and expansion of gas. This will reduce expensed and increase conducted work. Practically, approaching of curves will happen. Rates of compression and expansion of gas will also decrease to infinitely low values with the size of loads, decreased to infinitely low values, and time of conduction of both processes between points 1 and 4 rises up to infinitely large values. Under border conditions, compression and expansion of gas in this method of the process conduction will proceed by the same curve II. Such process can be performed only under infinitely slow conduction of it, when the number of points on the curve is infinitely great, and steps between equilibrium states of the system (points on the curve), which determine indefiniteness in the system state, will be infinitely small.

Consequently, if compression and expansion of gas are quasi-equilibrium, a non-spontaneous process is completely compensated by a spontaneous one, and works of expansion and compression of gas will be equal to each other. In this case, the work is also completely compensated by heat.

Processes proceeding at infinitely low rate, infinitely long, through the same states of equilibrium, direct and opposite, are called quasi-equilibrium (nearly equilibrium), quasi-static or for more simplicity – simply equilibrium ones. No changes happen in the surrounding in the equilibrium process, and such processes are called reversible. Processes of friction between substances and expansion of gas into empty space, transition of heat from hot substance to cold one at a finite difference of temperatures are thermodynamically irreversible.

In reversible processes of gas compression, square under the curve of non-spontaneous process will be minimal, and under the expansion curve – maximal. Consequently, the work of reversible process of compression is minimal, and of gas expansion – maximal, comparing with works of irreversible processes, and they are equal, i.e.

$$W^{max}_{expans.} = W^{min}_{compr.}. \qquad (4.14)$$

These values of works are extreme.

3. From the above-mentioned proof, it follows that equilibrium (thermodynamically reversible) processes are border states or conditions of conduction of real processes, which cannot be practically performed (equilibrium proceeding of the process) in reality. However, this notion is a suitable abstraction in thermodynamics, because it simplifies structure and composition of thermodynamic equations.

In conclusion, it can be notified that analytical expressions of the first law of thermodynamics can be applied to description of both reversible and irreversible processes, but quantitative calculations are suitable to be performed for equilibrium processes.

The second law will also be applicable to both types of processes, reversible and irreversible, but equations for irreversible ones are more complex and must take into calculation interactions of the system with the surrounding. In conclusion, let us note once again that subdivision of processes into spontaneous and non-spontaneous ones is possible for irreversible processes only.

4.5. Formulation of the Second Law of Thermodynamics

1. The second law of thermodynamics was formulated in order to determine conditions, under which thermal machines could, on one hand, be the source of work, and on the other hand, in order to determine impossibility of creation of *perpetuum mobile* of the second kind. Preliminarily, it can be noted that all formulations of the second law of thermodynamics reflect proceeding of thermodynamically irreversible processes. These formulations separate conditions of preferable proceeding of spontaneous processes over non-spontaneous ones.

More strictly, it can be notified that formulations of the second law of thermodynamics separate conditions, in which exceeding of a part of spontaneous processes over non-spontaneous ones is displayed, and vice versa. This is bound to the fact that both processes – spontaneous and non-spontaneous – are interconnected and proceed simultaneously and continuously.

Formulations of the second law of thermodynamics cannot relate to thermodynamics of equilibrium processes. Prevalence of spontaneous processes over non-spontaneous ones cannot be noted for equilibrium processes. Equilibrium processes proceed with infinitely low deviations from the equilibrium state of the system, at high degree of compensation of proceeding processes. Uncontrolled small changes to one or another side in parameters can cause proceeding of spontaneous or non-spontaneous processes. In equilibrium processes, the type of the process proceeding cannot be predicted because of the effect of accidental factors. Strictly speaking, formulations of the second law of thermodynamics for equilibrium processes are inapplicable.

Thermodynamically irreversible processes proceed at a finite rate, in which rates of spontaneous process υ_{sp} and non-spontaneous one $\upsilon_{n\text{-}sp}$ can be easily determined. Difference in these rates determines the rate of the resulting process. The rates of resulting processes are those, reflected in unclear form in formulae of the second law of thermodynamics, shown below.

2. R. Clausius has formulated the second law of thermodynamics as follows: *heat can transit spontaneously from hot substance to cold one and cannot transit spontaneously from cold to hot substance, without expense of work or any other type of compensation.* According to K.A.Putilov, compensations can be of two types. The first type of compensation is bound to change of the state of the working substance at heat transformation into work (the term of heat transformation into work is of the sense of one type process transfer into another). For example, at heating of the ideal gas in a cylinder with a plunger, the appropriate work is conducted by this gas, and increase of the gas volume happens as the compensation.

3. Compensation of the second kind is displayed in the case, when three substances — heater, working substance and cooler — participate in the cycle of the thermal machine operation (Figure 4.2).

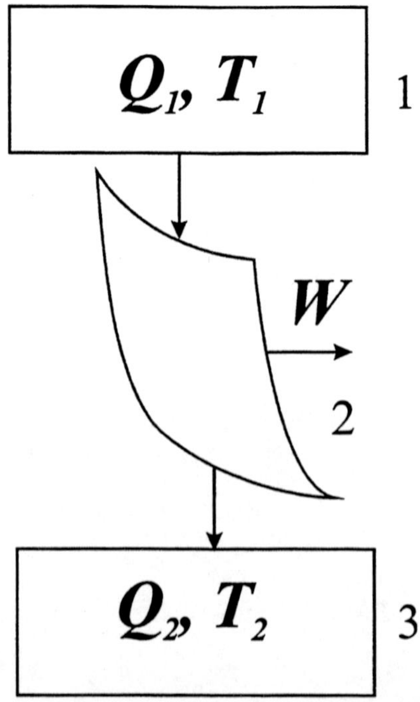

Figure 4.2. The complex of working substances in a thermal machine: 1 — heater; 2 — working substance; 3 — cooler.

In this cycle, the working substance is regularly recurred to the initial state, change of the cooler state plays the role of compensation of the second kind. In this thermal machine the working substance expands at one temperature, and a part of heat of the heater, Q_1, transforms into work, and the rest of heat, Q_2, is transmitted to the cooler. The amount of work conducted by the thermal machine will be

$$W = Q_1 - Q_2. \tag{4.15}$$

Efficiency of heat transformation into work for the thermal machine is determined by the efficiency coefficient (EC, η):

$$\eta = \frac{Q_1 - Q_2}{Q_1}. \tag{4.16}$$

On the grounds of this formulation, it can be additionally determined that *proceeding of processes, the only result of which must be transformation of heat into work, is impossible in the thermal machine, if the state of the working substance is regularly recurred to the initial one.*

Thermal machine is the source of work at $T_1 > T_2$, i.e. it is a mechanism transforming heat into work. Other thermal machines are mechanisms consuming work and transforming it into heat. Such processes — transformation of Q_1 into W and W into Q_1 — are equally spread in nature. This is determined by the fact that conjugated spontaneous and non-spontaneous processes proceed continuous and simultaneously in nature, amounts of which are equal, and this determines equal amount of two kinds (production and consumption) of work.

Systematic production of work by a thermal machine at the heat expense is impossible without compensation of the first or the second kind. This is reflected by the following formulation of the second law of thermodynamics.

Lord Kelvin (V. Thomson) has formulated the second law of thermodynamics as follows: *Heat of any substance cannot be transformed into work without performing any other action, except cooling this substance.* Consequently, thermal machine is able to produce work only under the condition that temperature of the heater is higher than that of the cooler. This is determined from the equation

$$\eta = \frac{T_1 - T_2}{T_1}, \tag{4.17}$$

where η (EC), T_1 and T_2 are the efficiency coefficient, temperatures of the heater and the cooler, respectively.

4. One more formulation states that heat of the working substance with temperature lower than that of the heater, cannot be the source of work.

V. Oswald has pointed out that a *perpetuum mobile* of the second kind cannot be constructed. This *perpetuum mobile* could transform heat into work completely, and such thermal machine would need no cooler, because EC of it would be equal to 1. It is also impossible to construct an engine working without difference in temperatures, i.e. under the condition of equality of temperatures

$$T_1 = T_2.$$

EC of this thermal machine will be equal to zero.

5. M. Plank has formulated the following definition of the second law of thermodynamics: *No regularly operating machine can be constructed, which would conduct only load lifting and cooling of the heat source.* It reflects the previous definition by V. Oswald.

Hence the above-mentioned formulations reflect the character of heat transformation into work, irreversible from thermodynamic grounds, and vice versa, work into heat. The source of work in this case is energy of a spontaneous process, and non-spontaneous ones may proceed only at the sacrifice of expense of work or excessive energy, produced by spontaneous processes. The latter reflect processes of accumulation of energy in the system or transmission of work to another system, etc.

Thermodynamics of equilibrium systems reflects possible, but inaccessible conditions of conduction of processes. This will be shown below in the form of the principle of entropy existence. This principle is the border theorem of real processes and reflects an idealized change of the system state. Fail of this principle, in terms of impossibility to reach conditions of its fulfillment in nature, excludes such notions as infinite lasting of objects in nature, immortality of biological objects, because changes in them proceed at a finite rate. Immortality supposes thermodynamically reversible proceeding of processes, i.e. at an infinitely low rate. The nature as it is, cannot permit immortality for its objects. The matter and forms of motion of it are immortal only, but these categories relate to philosophy. Existence of the matter is eternal with time and distance, in its continuous changing.

6. To describe quantitatively and understand spontaneous and non-spontaneous processes, a new thermodynamic function — entropy — is introduced. This parameter characterizes ability of a system to redistribute energy and substance in the current system or in interacting systems in the absence of external compensation of processes. Entropy determines the type and direction of changes of the system state in the case if spontaneous processes are compensated by non-spontaneous ones, especially when the latter prevail over the former. Spontaneous processes are suggested to be characterized by positive change of entropy, and non-spontaneous — by negative changes of entropy or neg-entropy. Growth of entropy in the current system or in a unity of systems reflects rise of homogeneous distribution of particles in the volume or growth of the chaos in the system, and decrease of entropy — rise of the order in the system. These two processes proceed simultaneously and continuously in systems, but the role of each of them in structure of the system depends on degree of the external influence on the system. If influence on the system is zero, the main role in it is played by spontaneous processes, in changing of the structure of it. At intensive influence on the system, neg-entropy and order in the structure and state of the system rise.

Entropy is the function of the system state. Changing of it does not depend on the way of the process proceeding. It is additive, one-valued and continuous function. One of drawbacks in understanding of this function is that it cannot be measured experimentally, with the help of a suitable instrument (chaosmeter). It can be calculated only.

7. M. Plank has formulated a postulate that formation of heat by friction is irreversible. This postulate, in fact, states impossibility to conduct processes by the equilibrium method under real conditions. Of importance is also the postulate by C. Carateodori, which determines that *near every equilibrium state of the system such states of it are possible, which cannot be reached by a reversible adiabatic process*. This postulate is just the one that determines impossibility of adiabatic transition from one closely disposed neighbor equilibrium state to another.

Moreover, this axiom points out the existence of an integrating divisor for a heat element, which transforms the equation of the first law of thermodynamics into the expression of the full differential or equation of the second law:

$$dS = \frac{dQ}{T}. \tag{4.18}$$

8. N.I. Belokon' has united postulates about incompatibility of adiabatic and isothermal curves, the Carnot theorem, and the theorem of thermal equilibrium of contacting substances into the following postulate: *No simultaneous spontaneous* *transmission of heat in opposite directions is possible between substances or elements of substances not existing in the thermal equilibrium — from higher heated ones to lower heat, and vice versa*. However, the absence of a notion on a non-spontaneous process in this postulate makes its formulation somewhat indefinite. The point is that the process of heat transmission by the working substance with lower temperature (comparing with the heater) to the cooler is non-spontaneous, and heating up of the cooler by absorbed heat — spontaneous. The latter is accompanied by rise of entropy. It is the decelerating one and characterizes possibility of temperature leveling in contacting substances at the sacrifice of transmission and absorption of heat. The rate of heat transmission and absorption of heat is characterized by the equation:

$$\frac{dQ}{d\tau} = \frac{dQ_{transm.}}{d\tau} - \frac{dQ_{absorp.}}{d\tau}. \tag{4.19}$$

Here $Q_{transm.}$ and $Q_{absorp.}$ are transmitted and absorbed heats, respectively.

Deceleration of the heat transfer can be hardly imagined without effect of the heater and cooler; moreover, equilibrium between the heater and the cooler under the condition $T_{heater} = T_{cooler} = T_{surround}$ cannot be reached without deceleration.

9. The second postulate by N.I. Belokon' maintains that *simultaneous conduction of complete transformations of heat into work and work into heat are impossible*. This postulate is true for thermodynamically irreversible processes, but this principle is also applicable to reversible ones, for which the law of equivalence is true. In reversibly conducted processes, as discussed in Section 4.4, the equalities are true:

$$W = Q \text{ and } Q = W. \tag{4.20}$$

Probably, this postulate can be more strictly formulated in the following view: *A gain in work of a thermal machine is impossible, if complete transformation of heat into work and work into heat is conducted simultaneously*. Such thermal machine works for itself under ideal conditions with additional energy expenses for friction (parts of mechanisms, molecular friction and other forms of energy scattering).

Hence it can be noted that the idea about equilibrium processes as the border condition of proceeding of real processes is suitable for application to composing simplified thermodynamic equations. These equations are easy to apply to calculation of the level of energy change in the system at its expense in the form of heat and work, with no consideration of mechanisms of the energy expense or consumption between two or more equilibrium states of the working substance. That is why, the principle of existence of entropy is the border postulate, which reflects idealized conditions of change in the state of the working substance and its interaction with the surrounding. This principle does not take into account continuous and different by sign fluctuations of the substance and energy in open systems. In closed, isolated systems, fluctuation is one of the mechanisms of approaching a non-equilibrium state of the system to an equilibrium one.

10. The principle of growth (decrease) of entropy for open systems reflects the real behavior of the working substance and its interaction with the surrounding. It can be added by the main postulate of kinetics — the total rate is proportional to the motive force, produced by difference of concentrations of substances in the system and amount of energy, the level of temperature, pressures, and other parameters, which reflect the state of interacting systems (working substances) and the surrounding.

Analytical expression of the second law of thermodynamics in the form of the new function — entropy, can be obtained on the basis of Carnot and Clausius theorems, and also with application of methods by R. Clausius, V. Thomson, N. Schiller—C. Carateodori and M. Plank. Theoretical proof of introduction of such a function into thermodynamics of equilibrium and irreversible processes does not exist yet. More simple way to approach the view of the new function is provided by Carnot—Clausius theorems.

4.6. Analytical Determination of Entropy

Analytical expression of entropy is obtained on the basis of V. Thomson postulates and S. Carnot and R. Clausius theorems, reflecting conditions of operation of thermal machines, which produce or consume work.

1. *V. Thomson postulate.* V. Thomson postulate reflects the real fact that cyclically operating thermal machine is the work producer, if the working substance participates in the round process between the heater and the cooler, heated up to different temperatures. Consequently, to transform heat into work, to transit from one process to another at transmission of energy in the heat form from hot substance to cold one, participation of three substances as minimum in all processes is required: heat source, working substance and cooler (heat acceptor) with corresponding assembly of spontaneous and non-spontaneous processes.

In this cycle, the working substance accepts heat from the heater in Q_1 amount. Heat transmission from the heater to the working substance represents a spontaneous process, and heat transformation into work — non-spontaneous one. In the ideal case, the working substance and the heater exist at the same temperature T_1. This condition is true for isothermal heating up of the working substance at infinitely small deviations of these two objects from equilibrium conditions. In this case, change of the state of the working substance — expansion, production of electric motive force and others — happens spontaneously, and transformation of a definite amount of energy into work is non-spontaneous process (taking into account adiabatic and isothermal curves of compression). Transmission of energy in the heat form in amount Q_2 and at temperature T_2 to heat acceptor or cooler happens during production of work by the working substance. In this case, the working substance is cooled down, and the cooler is heated up. Temperature of the cooler T_2 is lower than that of the working substance T_1, and this determines the motive force and direction of the energy transfer (heat elimination) in the thermal machine.

Difference in heats $Q_1 - Q_2$ is equal to the amount of work, produced by the thermal machine. Dividing this work by the amount of heat expensed Q_1, the efficiency coefficient of operating thermal machine is determined:

$$\mathrm{EC} = \frac{W}{Q_1}. \tag{4.21}$$

Analytical expression of entropy reflects the state of the working substance. To generalize spreading of this definition on any working substance, two theorems by S. Carnot and R. Clausius must be considered.

2. *The first Carnot-Clausius theorem.* The Carnot-Clausius theorem determines that *EC of a thermal machine, operating according to thermodynamically reversible cycle, does not depend on nature of the working substance and its state, but depends only on temperatures of heater and cooler.* This theorem is proved logically, on the basis of formulations of the second law of thermodynamics during discussion of operation of two conjugated thermal machines. One of these thermal machines operates in direction I and produces a definite amount of work, and the second machine II operates in the opposite direction and consumes work to release heat from the cooler, as shown in Figure 4.3.

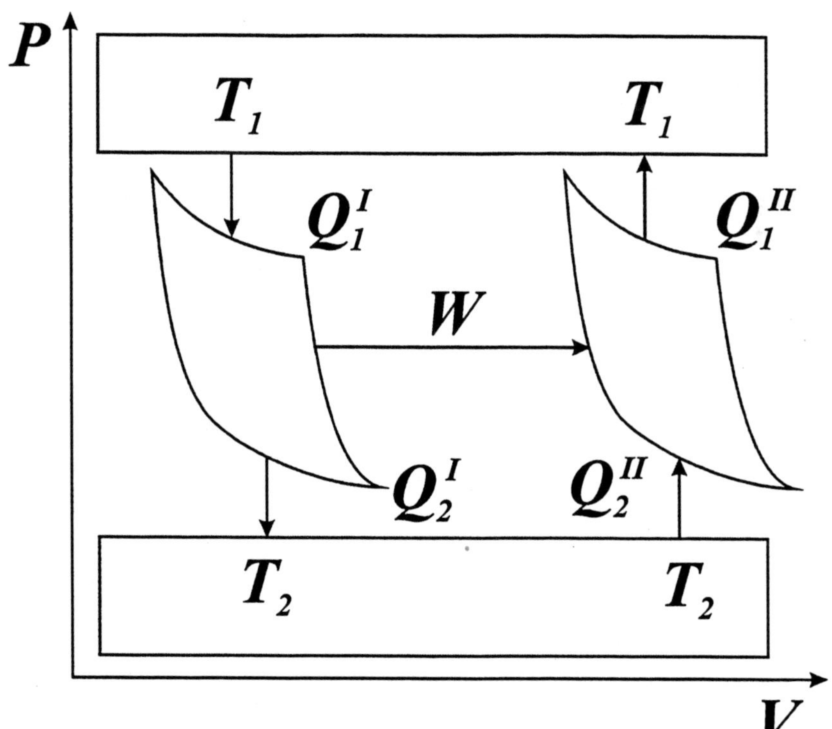

Figure 4.3. Scheme of operation of coupled thermal machines, operating with ideal gas (I) and real gas (II).

Let the machine I operate directly and produce work in the amount

$$W = Q_1^I - Q_2^I. \tag{4.22}$$

The efficiency coefficient of the thermal machine, operating directly, is expressed as follows:

$$\mathrm{EC}_{id.gas} = \frac{Q_1^I - Q_2^I}{Q_1^I}, \tag{4.23}$$

where Q_1^I and Q_2^I are amount of heat transferred from the heater to the working substance at temperature T_1, and amount of heat, transferred from the working substance to the cooler at temperature T_2.

Let the first (I) machine operate with the ideal gas, and the second (II) – with the real one.

Cooling machine II consumes the work W, at the sacrifice of which it releases heat from the cooler in amount of Q_2^{II} at temperature T_2 and transmits heat to the heater in amount of Q_1^{II} at temperature T_1.

Both machines operate according to reversible, ideal cycle of Carnot between the same heater and cooler. The efficiency coefficient of the cooling machine is determined by the equation

$$\mathrm{EC}_{real\ gas} = \frac{Q_1^{II} - Q_2^{II}}{Q_1^I}. \tag{4.24}$$

It can be assumed that EC of the thermal machine I is smaller than EC of operation of the cooling machine II, e.g.

$$\mathrm{EC}_{id.gas} < \mathrm{EC}_{real\ gas}. \tag{4.25}$$

Then we'll obtain the inequality

$$\mathrm{EC}_{id.gas} = \frac{Q_1^I - Q_2^I}{Q_1^I} < \frac{Q_1^{II} - Q_2^{II}}{Q_1^{II}} = \mathrm{EC}_{real\ gas}. \tag{4.26}$$

Impossibility of such inequality can be proved in the following manner.

In accordance with the law of conservation

$$\left(Q_1^I - Q_2^I\right) = \left(Q_2^{II} - Q_1^{II}\right). \tag{4.27}$$

As the machine operates conjugately, the work produced by the machine I equals to that consumed by the machine II:

$$W_I = W_{II}. \tag{4.28}$$

In accordance with different temperatures of the cooler and the heater, heats consumed and released by the working substance of direct and reverse machines are different:

$$Q_1^I > Q_2^I \text{ and } Q_2^{II} > Q_1^{II}. \tag{4.29}$$

The efficiency coefficient of the machine I is lower than EC of the machine II, that is why it consumes smaller amount of heat, than the cooling machine II transmits to the heater, i.e.

$$Q_1^{II} > Q_1^I, \tag{4.30}$$

and the heater will receive a non-compensated excess of heat in amount of

$$Q_1^{II} - Q_1^I = \Delta Q. \tag{4.31}$$

Total work of conjugated machines equals to zero according to the second law of thermodynamics:

$$W_I + W_{II} = 0. \tag{4.32}$$

Then under the above assumed admission, $EC_{id.gas}$ cannot be smaller than $EC_{real\ gas}$, the unique result of work of two conjugated machines of direct and reverse operation is the non-compensated transfer of heat from cold substance to heated one (spontaneous transfer of heat from cooler to heater), which is impossible according to the formulation by R. Clausius. Heat can transmit spontaneously from higher heated substances to less heated ones, whereas the reverse spontaneous transition of heat is impossible.

In can be assumed that EC of the cooling machine is lower than EC of the heating machine:

$$EC_{id.gas} > EC_{real\ gas}. \tag{4.33}$$

Then let us exchange the roles of machines, and, repeating all mentioned above, we'll obtain a contradiction with the second law of thermodynamics again, i.e.

$$EC_{id.gas} \not> EC_{real\ gas}. \tag{4.34}$$

The sign $\not>$ determines that EC of the heating machine cannot be smaller than EC of the cooling one.

According to formulation of the second law of thermodynamics, the equality of ECs of both machines remains true only:

$$EC_{id.gas} = EC_{real\ gas}. \tag{4.35}$$

Consequently, the efficiency coefficient of the heating machine, operating by the Carnot cycle, does not depend on nature of the working substance of the machine, but depends on temperatures of the heater and the cooler only (the Carnot-Clausius theorem). It must be noted that thermal machines can operate by irreversible cycles under real conditions, and by reversible Carnot cycles – in the border (not realized in reality) case. However, conclusions for both types of Carnot cycles – operation of heating machines – coincide. The same situation can be spread upon any cycles (finite by size, infinitely small by size, etc.).

The second Carnot theorem. The theorem by S. Carnot defines that: *the efficiency coefficient of the reversible Carnot cycle in the assigned range of temperatures of the heater and the cooler is greater than of any other arbitrary cycle of operating thermal machines*. Consequently, the form of the cycle defines EC value or efficiency of work of the thermal machine.

Analytical expression of the second law of thermodynamics enables work of the thermal machine to be determined by the expression:

$$W = Q_1 \frac{T_1 - T_2}{T_1}, \tag{4.36}$$

where T_1 and T_2 are temperatures of heater and cooler, respectively; W is the work, which can be determined by the expression, followed from consideration of the Carnot cycle:

$$W = Q_1 - Q_2. \tag{4.37}$$

3. Substituting (4.37) into (4.36), the expression for EC through the ratio of heats and temperatures is obtained:

$$\frac{Q_1 - Q_2}{Q_1} = \frac{T_1 - T_2}{T_1}. \tag{4.36}$$

Dividing the numerator by denominator, we obtain

$$1 - \frac{Q_2}{Q_1} = 1 - \frac{T_2}{T_1}.$$

The following correlations can be obtained from this equation:

$$\frac{Q_2}{Q_1}=\frac{T_2}{T_1}, \tag{4.39}$$

$$\frac{Q_1}{T_1}=\frac{Q_2}{T_2}, \tag{4.40}$$

where $\frac{Q}{T}$ is the reduced heat.

Equation (4.39) determines appearance of the temperature function:

$$\varphi(T_1,T_2)=\frac{T_2}{T_1}. \tag{4.41}$$

Correlation (4.40) is called the reduced heat. It displays that the ratio between heats of isothermal processes in the equilibrium transition of the working substance state from one adiabat to another and appropriate absolute temperature is the same for all isotherms and depends only on removal of adiabats from each other. Moreover, transition from one adiabatic curve to another is performed with the help of isothermal processes and cannot be performed adiabatically, i.e. without addition and elimination of heat to and from the working substance.

4. To prove the theorem, any round cycle can be used, inscribed into the ideal Carnot cycle, as shown in Figure 4.4.

The round cycle is characterized by equality to zero of the integral by closed contour from full differential of any function:

$$\oint dz=0. \tag{4.42}$$

The Carnot cycle is characterized by equality to zero of the sum of reduced heats (for finite or infinitely low values of isotherms and finite values of adiabats):

$$\frac{Q_1}{T_1}-\frac{Q_2}{T_2}=0 \tag{4.43}$$

or

$$\frac{dQ_1}{T_1}-\frac{dQ_2}{T_2}=0. \tag{4.44}$$

Considering two cycles and expressions (4.42)-(4.44) simultaneously, two separate tasks can be proved:

- EC of the Carnot cycle is greater than EC of an arbitrary round cycle;
- reduced heat equals to entropy, and its change does not depend on the way of the process proceeding.

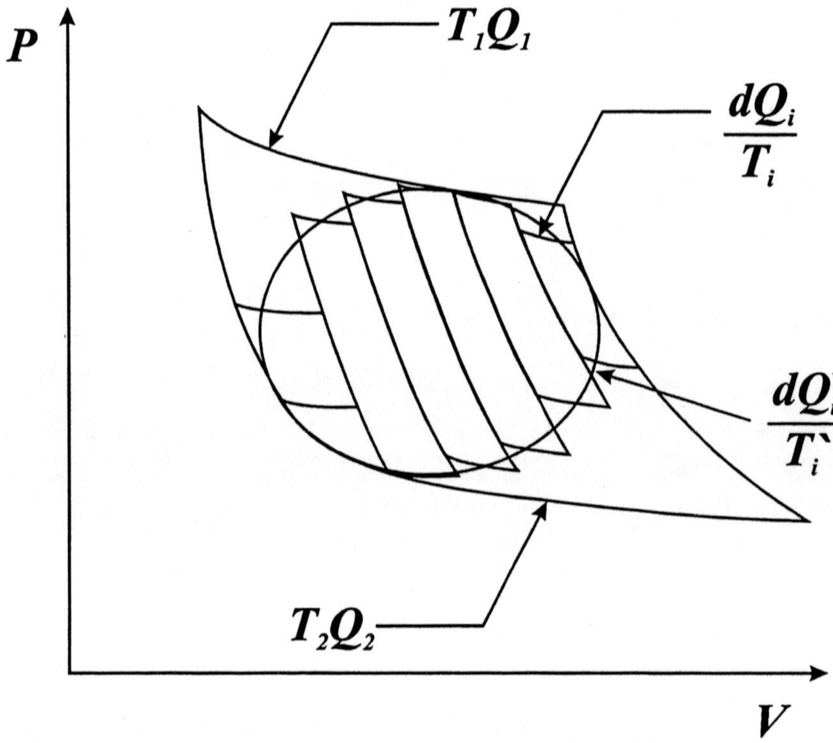

Figure 4.4. To the proof of the Carnot theorem and deduction of the analytical expression for entropy.

5. It can be observed from Figure 4.4 that the square of the Carnot cycle, equal to work W_{Cc}, is greater than the square of round arbitrary cycle W_{Cc} (where Cc marks the Carnot cycle), and from this point of view EC of the Carnot cycle is greater than that of the thermal machine, operating according to the round one. Analytically, this is proved as follows (K.A. Putilov). Let Q_1 and T_1, Q_2 and T_2 be heats and temperatures of the heater and the cooler, respectively, in the range of which thermal machine operates according to the ideal Carnot cycle. For the round cycle, it is necessary to determine average temperatures of heaters and coolers. For this purpose, let us divide the round cycle by isotherms of the extremely low value and finite adiabatic curves. Then the following sums of reduced heats can be composed for each low Carnot cycle:

$$\frac{dQ_1}{T_1} = \frac{dQ_1'}{T_1'},\ \frac{dQ_2}{T_2} = \frac{dQ_2'}{T_2'},\ \text{and so on.} \tag{4.45}$$

Comparing squares of the Carnot cycle and the round one, inequalities for EC of the cycles can be composed:

$$\frac{Q_1^{Ccycle} - Q_2^{Ccycle}}{Q_1^{Ccycle}} > \frac{Q_1^{r.cycle} - Q_2^{r.cycle}}{Q_1^{r.cycle}}. \tag{4.46}$$

For the round cycle, heats $Q^{r.cycle}$, produced by heaters and transmitted to coolers, can be presented in the form of equations of sums of low heats:

$$Q_1^{r.cycle} = dQ_1 + dQ_2 + dQ_3 + \ldots.., \tag{4.47}$$

$$Q_2^{r.cycle} = dQ_1' + dQ_2' + dQ_3' + \ldots.. . \tag{4.48}$$

Let us multiply and divide heats in the right part of these equations by temperatures of isotherms:

$$Q_1^{r.cycle} = T_1 \frac{dQ_1}{T_1} + T_2 \frac{dQ_2}{T_2} + \ldots.., \tag{4.49}$$

$$Q_2^{r.cycle} = T_1' \frac{dQ_1'}{T_1'} + T_2' \frac{dQ_2'}{T_2'} + \ldots.. . \tag{4.50}$$

Reduced heats are equal, if adiabats are disposed on equal distances from each other, i.e.

$$\frac{dQ_1}{T_1} = \frac{dQ_1'}{T_1'} = \frac{dQ_2}{T_2} = \frac{dQ_2'}{T_2'} = \ldots.. . \tag{4.51}$$

Substituting (4.51) into (4.49) and (4.50), values of heats for the totality of heaters and coolers for an arbitrary round cycle are obtained:

$$Q_1^{r.cycle} = \frac{Q_1}{T_1}(T_1 + T_2 + T_3 + \ldots..), \tag{4.52}$$

$$Q_2^{r.cycle} = \frac{Q_2}{T_2}(T_1' + T_2' + T_3' + \ldots..). \tag{4.53}$$

Average temperatures will be determined by the following expressions:

$$T_1^{av.r.cycle} = \frac{T_1 + T_2 + T_3 + \dots}{n}, \tag{4.54}$$

$$T_2^{av.r.cycle} = \frac{T_1' + T_2' + T_3' + \dots}{n}, \tag{4.55}$$

where n is the number of isotherms. Therefrom we'll obtain

$$Q_1^{r.cycle} = \frac{Q_1}{T_1} T_1^{av.}, \; Q_2^{r.cycle} = \frac{Q_2}{T_2} T_2^{av.}. \tag{4.56}$$

The efficiency coefficient for the thermal machine operating according to an arbitrary round cycle is determined from the formula:

$$EC_{r.cycle} = \frac{Q_1^{r.cycle} - Q_2^{r.cycle}}{Q_1^{r.cycle}}. \tag{4.57}$$

Substituting (4.56) into expression (4.57), we get:

$$EC_{r.cycle} = \frac{\frac{Q_1}{T_1} T_1^{av.} - \frac{Q_2}{T_2} T_2^{av.}}{\frac{Q_1}{T_1} T_1^{av.}}. \tag{4.58}$$

Contracting reduced heats due to the condition of equality of isotherms, we get a simple expression for EC of the round cycle:

$$EC_{r.cycle} = \frac{T_1^{av.} - T_2^{av.}}{T_1^{av.}}. \tag{4.59}$$

EC of the Carnot cycle will be determined by the formula:

$$EC_{Ccycle} = \frac{T_1 - T_2}{T_1}. \tag{4.60}$$

The average absolute temperature of the round cycle heater is lower than that of the Carnot cycle, and the average temperature of the cooler is higher than that of the Carnot cycle:

$$T_1^{av.} < T_1 \text{ and } T_2^{av.} < T_2. \tag{4.61}$$

Then the difference in temperatures for the round cycle will be smaller than for the Carnot cycle:

$$T_1^{av.} - T_2^{av.} < T_1 - T_2. \tag{4.62}$$

That is why

$$\frac{T_1^{av.} - T_2^{av.}}{T_1^{av.}} < \frac{T_1 - T_2}{T_1}. \tag{4.63}$$

Consequently, *EC* of the round cycle will be lower than that of the thermal machine operating according to the Carnot cycle:

$$EC_{r.cycle} < EC_{Ccycle}, \tag{4.64}$$

which is what we set out to prove. The left part of equation (4.63) is the generalization of various cycles and the Carnot cycle, in particular.

In approximation of values of absolute average temperatures of heaters for the round cycle by absolute temperatures of heaters of thermal machines operating according to the Carnot cycle the statement is true: the closer is the average absolute temperature of coolers to T_2, the closer are values of *EC* for two types of machines.

Basing on two cycles discussed above, two important results can be formulated:

- thermal machine operating according to the ideal Carnot cycle is analytically described by the sum of reduced heats;
- thermal machine operating by the round cycle is characterized by the expression

$$\oint dz = 0. \tag{4.65}$$

6. Comparing these two results, a new function – entropy – can be introduced into thermodynamics.

First of all, it should be noted that the round cycle equals to the sum of elementary reduced heats:

$$\frac{dQ_1}{T_1} - \frac{dQ_1'}{T_1'} + \frac{dQ_2}{T_2} - \frac{dQ_2'}{T_2'} + ... = 0. \tag{4.66}$$

This expression can be reduced to two sums – for upper and lower branches of the round cycle:

$$\sum \frac{dQ_i}{T_i} - \sum \frac{dQ'_i}{T'_i} = 0. \tag{4.67}$$

Transferring to the limit, the sum can be changed to integration by two directions:

$$\int_1^2 \frac{dQ_i}{T_i} - \int_2^1 \frac{dQ'_i}{T'_i} = 0. \tag{4.68}$$

Integration of reduced heats is conducted by two directions, leading to the initial state of the working substance. That is why two integrals can be substituted by an integral to the closed contour:

$$\oint \frac{dQ}{T} = 0. \tag{4.69}$$

This expression displays that the integral of elementary reduced heats by the closed contour does not depend on the way of the process proceeding, but depends on initial and final states of the system only. R. Clausius has selected a function, which fulfills this condition. According to the mathematical theorem that if an integral by closed contour from full differential of a variable equals to zero, there exists a function of integration variables, the full differential of which equals to the integrand value.

The integrand for the infinitely small increment of the reduced heat is presented as the full differential of the variable S:

$$dS = \frac{dQ}{T}. \tag{4.70}$$

In 1865, R. Clausius has named this function as entropy. The word “entropy” is composed of the preposition *en* – in, and the word *trope* – transformation, which means “transformation in /inside/”.

7. *Expression (4.70) reflects the principle of existence of entropy, i.e. the measure of order in the system.* Systems with high order and more regular structure possess lower entropy. Energetically, entropy is considered as the measure of invaluable energy at constant temperature:

$$dQ = TdS. \tag{4.71}$$

Systems with growing entropy are characterized by less ordered structure of the working substance. These two conclusions were formulated for the equilibrium change of state of the system or the working substance interacting with the surrounding. As follows from equation (4.69), entropy of the working substance is the one-valued, continuous and finite function of the system state, and its change does not depend on the way of process proceeding. This function is extensive. Taking into account expressions (4.69) and (4.70), it follows that at the end of the round cycle entropy obtains the initial value. Then the integral by the closed contour from the full differential of entropy will be equal to zero:

$$\oint dS = 0. \tag{4.72}$$

It must be noted that the equality sign in expressions (4.63)-(4.72) reflects thermodynamically reversible processes, which cannot be realized under real conditions, and is a suitable abstraction for calculation of changes of thermodynamic functions. Thermodynamically reversible processes must proceed at the infinitely low rate, i.e. infinitely slowly and during infinitely long time. Equations (4.59)-(4.72) reflect the principle of entropy existence, i.e. the ability of the working substance or system to change the structure, composition and state in a thermodynamically reversible way.

4.7. Temperature Scale as the Level of Entropy Reading

1. Temperature determines one of properties of the system, bound to energy of it, its store and changes during various processes (heat transfer, work, changes of structure of the working substance, aggregate state, etc.). Together with mass, time and size, temperature is the main parameter of the system.

The value of temperature changes from -∞ through 0 to +∞. Scaling of temperature is performed with the help of molecular-kinetic and thermodynamic methods.

According to the molecular-kinetic theory, temperature of the working substance is bound to kinetic energy of moving molecules, and temperature is the measure of energy.

Equation, which enables temperature to be determined, is composed on the following basis:

- the main equation of molecular-kinetic energy:

$$PV = \frac{1}{3} m\upsilon^2, \tag{4.73}$$

- equation of the kinetic energy:

$$E = \frac{1}{2} m\upsilon^2, \tag{4.74}$$

- and the Maxwell-Clapeyron equation:

$$PV = RT, \tag{4.75}$$

where υ is the mean rate of molecule transposition in the system, m is the mass of molecules per one gram-mole, E is energy, P, V are pressure and volume, respectively, R is the universal gas constant. Substituting (4.75) into (4.73), we obtain:

$$RT = \frac{1}{3} m\upsilon^2. \tag{4.76}$$

Substitution of (4.74) into (4.76) gives:

$$RT = \frac{2}{3} E. \tag{4.77}$$

Therefrom temperature is connected with energy of the working substance:

$$T = \frac{2}{3} E \frac{1}{R}. \tag{4.78}$$

It can be noted from this equation that the initial value of temperature is characterized by equality of energy to zero:

$$E = 0. \tag{4.79}$$

Under these conditions, the absolute temperature equals to zero:

$$T = 0. \tag{4.80}$$

Equation (4.79) is physically senseless, because is suggests disappearing of energy at the absolute zero, i.e. its transformation into nothing. Really, energy is connected with the substance and field properties of the matter, with various forms of the matter motion and carriers of the matter motion. Carriers of motion of the matter and motion of the matter itself are indestructible. That is why equation (4.79) supposes disappearance of carriers of motion in company with the motion, which is impossible from grounds of natural science and properties of the matter. Energy can be only transmitted from one carrier to another and transform from one type into another.

That is why, if equation (4.80) reflects the border state of the system, equation (4.79) reflects a possibility of disappearance of one of the forms of the matter motion – molecular motion, and does not relate to other forms of motion of the matter, which cannot be equal to zero. In fact, this reflects only a theoretical possibility of reaching the zero value of the absolute temperature by the system. In reality, impossibility to stop moving by material objects is reflected in impossibility of reaching the zero value of temperature under any effects on the working substance.

The Mendeleev-Clapeyron and Gey-Lussacs-Charles equations display that temperature is in linear dependence on volume and pressure of the working substance. This property was put to the basis of creation of gas thermometers. Other properties – thermoelectric, calorimetric, and others – can also change linearly with temperature.

2. From thermodynamic grounds, the notion of temperature is introduced on the basis of formulations of the second law of thermodynamics, analytical expressions of the second law of thermodynamics and Carnot-Clausius theorems.

The second law of thermodynamics in the form of the efficiency coefficient determines correlations between heats of the heater and the cooler by equation of the following type:

$$\frac{Q_1 - Q_2}{Q_1} = \frac{T_1 - T_2}{T_1}. \tag{4.81}$$

According to the first law of thermodynamics, difference in heats of the heater Q_1 and the cooler Q_2 equals to the work of the thermal machine W, operating by the ideal Carnot cycle:

$$W = (T_1 - T_2)\ln\frac{V_2}{V_1}, \tag{4.82}$$

$$W = Q_1 - Q_2, \tag{4.83}$$

where V_1 and V_2 are the initial and the final volumes of the working substance, respectively; T_1 and T_2 are temperatures of the heater and the cooler.

As follows from formulae (4.82) and (4.83), the work produced by a thermal machine is proportional to the difference in temperatures of its heater and cooler. The amount of heat transformed into work is determined by the efficiency coefficient of the thermal machine, and it is proportional to the ratio of temperature difference to the heater temperature.

Substituting (4.83) into (4.81), it is obtained that

$$\frac{W}{Q_1} = \frac{\Delta T}{T_1},\ W = Q_1\frac{\Delta T}{T_1}, \tag{4.84}$$

relative difference in temperatures is the measure of heat transformation into work. This value does not depend on origin of the working substance or the type of energy – the source of work and, consequently, is the universal law of temperature determination.

3. We can easily come to universal determination of temperature basing on equation (4.81). Dividing numerator by denominator in this equation, the following expression is obtained:

$$1-\frac{Q_1}{Q_2}=1-\frac{T_1}{T_2} \text{ or } \frac{Q_1}{Q_2}=\frac{T_1}{T_2}. \tag{4.85}$$

This equation displays that heats and temperatures are connected by linear correlations:

$$Q_2 = aT_2, \tag{4.86}$$

$$Q_1 = aT_1, \tag{4.87}$$

where a is the scale coefficient.

Temperature can be both positive and negative and changes from 0 to +∞ or from 0 to -∞. These correlations do not clearly reflect the nature and structure of the working substance, and the scale coefficient depends on numerical values of reference points or calibration grades of temperatures (centigrade, Reaumur, Fahrenheit, Kelvin, etc.). The most simple temperature scale is obtained under the condition that coefficient a equals to 1.

4. Temperature and bound energy. Linked energy is determined from the product of constant temperature and entropy:

$$E = TS. \tag{4.88}$$

This energy cannot be transformed into work under constant temperature. It follows from this expression that temperature equals to the ratio of linked energy and entropy:

$$T=\frac{E}{S}. \tag{4.89}$$

The following features can be noted from equation (4.89):

1) *entropy cannot be equal to zero at the absolute zero of temperature only, because in this case expression (4.89) strives to the infinity at $E \neq 0$ and $S = 0$ or becomes an indeterminacy at $E = 0$ and $S = 0$.* Evidently, if $E \neq 0$, it appears

$$T=\frac{E}{0}\rightarrow\infty. \tag{4.90}$$

If $E = 0$ and $S = 0$, we get

$$T = \frac{0}{0}. \tag{4.91}$$

That is why the condition $T = 0$ K requires the following condition for state of the working substance to be fulfilled near the absolute zero:

$E = 0$, $S \neq 0$ and $T = 0$.

Then

$$0 \times S = 0. \tag{4.92}$$

Condition (4.92) determines that entropy of the working substance cannot be equal to zero even at the absolute zero.

2) *To scale temperature, one can base on the condition, independent on nature of thermometric substance and its structure*:

$$E = S. \tag{4.93}$$

This condition is strict at $T = 1°$. Hence, the standard scaled unit of temperature corresponds to the interval of the working substance state from 0 to 1°, consequently,

$$\Delta T = 1°. \tag{4.94}$$

If the possibility of approaching the absolute zero from both sides, positive and negative, is considered, equality (4.94) will be reduced to the following view:

$$\Delta T = \pm 1°. \tag{4.95}$$

The minus sign relates to spontaneous decrease of entropy, and the plus sign – to spontaneous rise of entropy for the working substance.

Temperature scale can be determined by radiation of the working substance applying the Stephan-Boltzmann law and the Carnot cycle, as well as the magnetic scale of temperatures can be applied with the Curie law and the Carnot cycle. All these temperature scales coincide.

5. Statistic ground of temperature. Statistic ground of entropy follows from the Boltzmann equation:

$$S = \frac{R}{T} \ln \omega. \tag{4.96}$$

From this expression temperature is determined by the formula:

$$T = \frac{R\ln\omega}{S}, \tag{4.97}$$

$$\omega = \frac{N!}{N_1!N_2!...}, \tag{4.98}$$

where $N!$ is the number of transpositions of N molecules, $N_1!$, $N_2!$, ... are numbers of transpositions of molecules in the range of the 1st, 2nd, and so on cells, ω is thermodynamic probability of state of the system containing N molecules and n cells (1, 2, 3, ...). From the point of view of statistics, temperature characterizes the degree of ordering of the system structure. Temperature can become zero only under the condition that $\omega = 1$, and $S \neq 0$.

Consequently, according to expression (4.97), entropy must become zero somewhat later than temperature and logarithm of thermodynamic probability. Otherwise temperature will become not zero, but the infinity will occur. This fact must be understood. Evidently, when temperature decreases, processes leading to increasingly significant ordering of disposition of particles at long distance relatively to each other proceed in structure of the working substance. Any ordering of particles terminates near the absolute zero, i.e. in its neighborhood. This is determined by constancy of working substance properties in the neighborhood of the absolute zero, and the working substance adheres a structure, strictly predicted from the point of view of thermodynamics and is expressed by $\omega = 1$. That is why a state of the working substance is obtained, which is yet characterized by border, non-zero value of entropy, but constant temperature, which can be determined as the absolute zero. From these grounds, the zero value of temperature is the scientific abstraction, impossible to reach and strictly determine. The zero value of entropy can be reached by extrapolation of properties of the working substance to zero only.

4.8. Generalized or the Main Equation of Thermodynamics for Equilibrium and Non-Equilibrium Processes.

1. Generalized or the main equation of thermodynamics reflects heat and entropy processes and contains analytical expressions of the first and the second law of thermodynamics. Actually, equation of the first law of thermodynamics is more general, than that of the second law and its left part. The equation of the first law of thermodynamics acquires a particular form, when entropy is introduced into it.

For equilibrium processes, the equation of the first law of thermodynamics is written in the following form:

$$dQ_{eq} = dU_{eq} + dW_{eq}, \tag{4.99}$$

where *eq* marks the equilibrium. It displays that heat of the equilibrium process equals to the internal energy and work, the border values of non-equilibrium processes. This equation can be hardly understood, because it does not reflect really proceeding processes. For equilibrium processes, the notions of *absorption* or *release* of energy in the heat form for current working substance, participating in conduction of an equilibrium process, is conceptually senseless due to the definition of equilibrium processes themselves. An infinitely small deviation in behavior of the working substance from the state of equilibrium to one or another side happens in equilibrium processes. In this case, the sign of these deviations cannot be determined, because they happen randomly.

Real processes proceed with a finite deviation from the state of equilibrium in their properties, and the value of this deviation determines direction and rate of the process proceeding. Formulations of the first law of thermodynamics reflect these deviations and irreversibility of the process proceeding, but not in equation (4.99). Processes in current system, interacting with the surrounding, can proceed spontaneously and non-spontaneously, with release or absorption of energy, work and heat.

2. The system can be heated up by addition of heat or work to it. Such processes are called non-spontaneous one. The system releases energy in the heat form into the surrounding with cooling down itself. The latter process is spontaneous one. Heat release of the present system into the surrounding is the spontaneous process, and transformation of the internal energy into heat and work — non-spontaneous.

Subdivision of processes into spontaneous and non-spontaneous ones, which are exchanged between the system and the surrounding, is possible for thermodynamically irreversible processes only.

3. The first law of thermodynamics is formulated for thermodynamically irreversible processes as follows:

- *heat added from the surrounding to the system (working substance or unity of substances) is expensed for rise of the internal energy and production of work by this system*;
- *heat released from the system into the surrounding is expensed for rise of the internal energy and production of work in the surrounding.*

These two formulations reflect different analytical equations of the first law of thermodynamics. For the sake of convenience of composing equations for the mentioned formulations, let us agree that the system reflects the internal part of the considered object in composition of the system and the surrounding. The system is marked by letter *i* — internal. The surrounding, which contacts with the present system, is determined as the external part of the object and is marked by letter *e* — external.

Applying these designations, equations of the first law of thermodynamics will be written in the form, different from the above present one:

$$dQ_e = dU_i + dW_i, \qquad (4.100)$$

$$dQ_i = dU_e + dW_e. \tag{4.101}$$

These equations can be integrated in the range from the first to the second state and vice versa.

This separation of equations is conditional, because two processes proceed continuously and simultaneously in the unity of the system and the surrounding. Basing on equations (4.100) and (4.101), the general equation can be obtained for the unity of the system and the surrounding:

$$dQ_e + dQ_i = dU_e + dW_e + dU_i + dW_i \tag{4.102}$$

or

$$dQ = dU + dW. \tag{4.103}$$

Heat in the left part of equations (4.102) and (4.103) can be substituted according to the principle of existence of entropy by the equation of the following view:

$$dS = \frac{dQ}{T}, \tag{4.104}$$

or for each type of processes the principle of existence of entropy will be determined separately by expressions:

$$\frac{dQ_e}{T_e} = dS_e, \tag{4.105}$$

$$\frac{dQ_i}{T_i} = dS_i. \tag{4.106}$$

These equations reflect indirectly principles of thermodynamics of irreversible processes, but this statement will be proved later.

Substituting equations (4.105) and (4.106) into (4.102), we get the main equation of thermodynamics:

$$dS_i + dS_e = \frac{dU_i + dW_i}{T_i} + \frac{dU_e + dW_e}{T_e}, \tag{4.107}$$

or grouping terms, determining the internal energy and work, reduce to the following expression:

$$dS = \left(\frac{dU_i}{T_i} + \frac{dU_e}{T_e}\right) + \left(\frac{dW_i}{T_i} + \frac{dW_e}{T_e}\right), \tag{4.108}$$

where $dS = dS_i + dS_e$.

Equation (4.108) can be rewritten for the totality of phase-separated subsystems inside the present system with different values of temperature and heat, which interact with the surrounding, as follows:

$$dS = \frac{dQ_e}{T_e} + \sum \frac{dQ_i}{T_i} - \sum \frac{dQ_j}{T_j}, \tag{4.109}$$

where dQ_i, T_i are heat and temperature of heated up parts of the system; dQ_j, T_j are heat and temperature corresponded to cold parts of the system. Heat redistribution proceeds between these parts of the system and the surrounding.

Assuming that heats are equal:

$$dQ_i = dQ_j, \tag{4.110}$$

we transit to the main equation of thermodynamics as follows:

$$dS = \frac{dQ_e}{T_e} + \sum dQ_i\left(\frac{1}{T_i} - \frac{1}{T_j}\right). \tag{4.111}$$

The general view of equation (4.111) reflects a change of entropy inside the current system and the surrounding, with which the system interacts via totality of subsystems. This equation reflects proceeding of spontaneous and non-spontaneous processes in interacting system and the surrounding.

4. Other expressions for entropy of the system and the surrounding in unity of their interaction can be deduced. Summing up expressions (4.105) and (4.106), we obtain:

$$dS_e + dS_i = \frac{dQ_e}{T_e} + \sum \frac{\delta Q_i}{T_i}, \tag{4.112}$$

i.e. the system and the surrounding exchange the reduced heat continuously, changing dS_e and dS_i entropy values.

Multiplying entropy by temperature in equations (4.105) and (4.106), the sum of heats will be characterized yet by other functions of state of the system:

$$dQ_e + dQ_i = T_e dS_e + \sum T_i dS_i. \tag{4.113}$$

In this equation, intensive parameters are temperatures of the surrounding T_e and the system T_i. Prevalence of the type of process —spontaneous or non-spontaneous — in the above-mentioned system of equations can be determined after theorem is proved, discussed in the next Chapter.

Chapter 5

THERMODYNAMICS OF SPONTANEOUS AND NON-SPONTANEOUS PROCESSES

5.1. NECESSITY OF DEVELOPMENT OF A NEW DIRECTION IN THERMODYNAMICS

Until the present time, the attention paid by investigators to the interaction of spontaneous and non-spontaneous processes in nature and technology was insufficient. This is connected with the initial stage of development of thermodynamics of irreversible processes — thermodynamics of spontaneous, stationary and non-stationary processes. Thermodynamics of irreversible processes studies thermal processes and the ones of transfer and transformation of substances in spontaneous processes.

Heat transmission from heated substance to cold one is the spontaneous process, and cooling of the working substance down to a state below the equilibrium one is the non-spontaneous process. Both these processes proceed in a thermal machine. Thermal machine operates with three substances – heater, working substance and cooler. The heater releases heat to the working substance, and this process is spontaneous for the heater. Receiving a definite amount of heat, the working substance changes internal parameters. This process is spontaneous (heating of the working substance). The working substance transforms a definite amount of heat into work, and this process is already the non-spontaneous one.

Heat that does not participate in work is transmitted to the cooler. This process is non-spontaneous, at the sacrifice of this heat the cooler is heated up, and this process is spontaneous. The work is consumed for friction, which appears non-spontaneously at the sacrifice of forces of interaction between friction surfaces and transforms into heat – the spontaneous process. Other pairs of spontaneous and non-spontaneous processes can be also separated in this system, which will allow deeper understanding of mechanisms of the thermal machine operation and separation of general regularities of their work on the basis of particular regularities of interaction of a pair of such processes.

Regularities of interaction of spontaneous and non-spontaneous processes must be set up for various physical, chemical and physicochemical processes in order to clear up fine mechanisms of diffusion and mass transfer, chemical transformation of substances and combination of physical processes of diffusion and mass transfer with chemical processes. The same relates to mechanical and biological forms of the matter motion.

5.2. Spontaneous and Non-Spontaneous Processes on Different Levels of State of the Matter

1. In accordance with ideas of natural sciences, discussed in Section 1.4, state and complex forming of structure of the matter and its material composition can be presented by a simplified scheme (Figure 5.1).

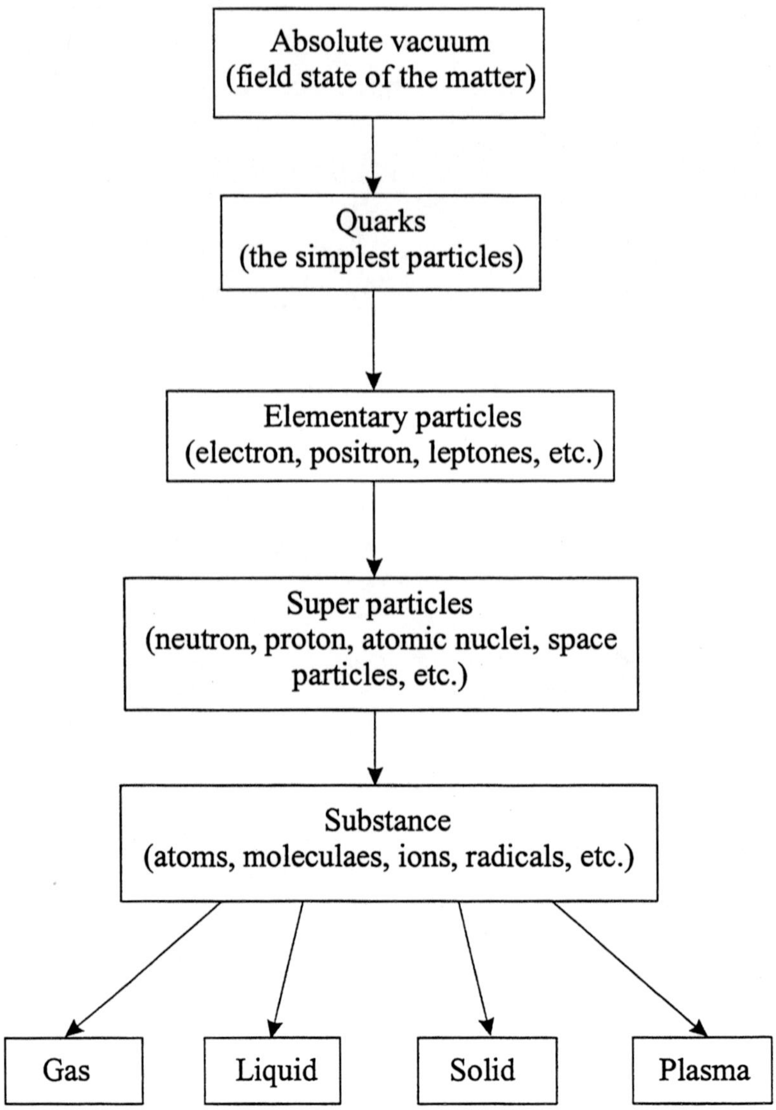

Figure 5.1. Scheme of leveling of state of the matter and its material composition.

2. The absolute vacuum represents a field complicated in composition, structure and state, which are yet studied insufficiently. The absolute field is the source of occurrence of quarks and receiver of the field state of quarks. The process of formation of quarks from turbulences and fluctuations of the absolute field proceeds spontaneously with participation of the field form of the matter motion. Transition of quark in the state of the absolute vacuum is the non-spontaneous process and requires expenses of a tremendous amount of energy. Hence, a pair of processes – spontaneous and non-spontaneous – is one of initial stages of transformation of the material state and structure of the matter. The same pair of processes is one of the finite transformation stages of the material state of the matter into the absolute vacuum. In the surrounding, no other sources of formation of the material state and structure of the matter and its field state can be separated.

3. Uniting with each other, quarks form elementary particles, the most important of which for formation of the matter are electrons, positrons, protons and neutrons. Synthesis of these elementary particles proceeds spontaneously, with release of energy, and reverse processes proceed non-spontaneously, with expense of a tremendous amounts of energy. Each of stages of synthesis or decomposition of elementary particles may include a significant number of stages, on which influence corresponding pairs of spontaneous and non-spontaneous conjugated, dependent and continuously proceeding processes.

Elementary particles and super-particles combine with each other forming a substance existing in the form of four aggregate states – plasma, gas, liquid, and solid.

One of directions of change of the absolute vacuum state is its transition into anti-quarks, anti-particles and anti-substance. In its turn, substance and anti-substance can exist under the conditions of positive and negative absolute temperature. Under these conditions, spontaneous and non-spontaneous processes exchange places. Spontaneously proceeding processes under conditions of negative temperatures become non-spontaneous in the range of positive temperatures and vice versa.

The notion of anti-matter, anti-particle and anti-substances is complex. The notion of anti-temperature must be introduced for these states and structures of the matter, which is not yet understandable from the physical point of view.

5.3. Phase-Open Systems

1. The surrounding is a unity of phase-open systems. The systems are called phase-open, if separated from each other by a phase boundary, possess equal physicochemical and thermodynamic properties and composition, and interact with each other exchanging energy and substance. Various objects of the surrounding can be the phase-open systems.

2. The mentioned spontaneous and non-spontaneous processes proceed simultaneously and continuously in conjunction of phase-open systems or a system and the surrounding. Spontaneous processes prevail over non-spontaneous ones in some systems, and such systems degrade. In other systems, non-spontaneous processes prevail over spontaneous ones, and such systems are developing. The same processes also proceed at interaction of phase-open systems with the surrounding.

A phase-open system is of current volume and content, but their number is not defined in the environment. The environment represents the infinity in time and space. The number of phase-open systems in the environment is infinite, and the number of pair of interacting spontaneous and non-spontaneous processes is also infinite.

Infinity of the environment (the Universe) reflects reality of its existence, without necessity of external influence to change the beginning and end on existence of it. Definite changes in the World can be noted for conditionally chosen temporal space only from the point of view of its local part of the type of the Solar System, galaxies, the Universe as a part of the Para-Universe. For systems and the surrounding, these changes are characterized by interactions of spontaneous and non-spontaneous processes.

3. Two types of development of a phase-open system and the surrounding can be selected. The phase-open system can possess an excess of free energy at higher values of intensive parameters and lower value of entropy, than the surrounding. Then the excess of free energy will transit spontaneously from the phase-open system into the surrounding. In this case, the store of energy in the surrounding grows with the entropy value. The excess of free energy and the value of entropy decrease in the system, and the processes proceed up to leveling of intensive parameters. Consequently, spontaneous and non-spontaneous processes in the interacting complex of objects can be separated as follows:

System

Decrease of the free energy excess	**- spontaneous process**
(T_i, P_i, ... > T_e, P_e, ...)	
Cooling down of the system ($T_i < T_e$)	**- non-spontaneous process**
Decrease of entropy ($T_i < T_e$)	**- non-spontaneous process**

The surrounding (or a conjugated system)

Growth of the energy excess ($T_e > T_i$)	**- non-spontaneous process**
Rise of temperature ($T_e > T_i$)	**- non-spontaneous process**
Growth of entropy ($T_e > T_i$)	**- spontaneous process**

Consequently, if the system possesses higher potential possibility to transmit energy into the surrounding, than vice versa, the main process in this complex of objects will be a spontaneous one.

Proceeding of the spontaneous process will be decelerated by non-spontaneous ones, which proceed in conjugation with spontaneous ones.

4. Phase-open systems may be characterized by lower store of potential (free) energy, than the surrounding or another system, conjugated to the former one, or a combination of systems. Then the excess of potential, or free, energy will be transmitted from the surrounding to a phase-open system on account of its higher heating or owing to the work consumption for energy transfer from one substance to another. For these purposes, other kinds of sources of the excessive energy can be used – electric energy, chemical energy, biological energy, electromagnetic energy, and others. A non-spontaneous process will prevail in such combinations of objects. Complexity of these processes, continuously and with a conjugation proceeding under interaction with the surrounding, is the property of the matter and its material state, and does not depend on the spatial-temporal factors from the point of view of the matter existence and forms of its motion. From this point of view, ideas on creation of the World (explosions of compressed matter, the Lord contribution, and other methods) are far from the natural science theory and must be based on the above-mentioned scientific grounds only, which reflect the main laws of development of the matter and its field and material state.

5. From the foregoing it can be noted that the main laws of thermodynamics – zero, first, second, and third – require formulation of the fourth law of thermodynamics, which reflects proceeding of interacting spontaneous and non-spontaneous processes in phase-open systems and the surrounding.

The law can be formulated as follows: *conjugated spontaneous and non-spontaneous processes proceed in the conjunction of phase-open systems and the surrounding, and spontaneous processes proceed in the system with decrease of the free energy excess and entropy, and in the surrounding conjugated non-spontaneous processes proceed with rise of the free energy and entropy.* This formulation reflects processes proceeding in interacting phase-open systems and the surrounding in more detail, than it has been formulated before in the book "Thermodynamics of physicochemical processes" by I.M. Kolesnikov, Moscow, 1994 (Rus). In this book, the formulation reflects processes proceeding in the range of a single phase-open system and the surrounding.

6. This law was formulated in the form of a theorem, which can be proved on the basis of logical reasoning with application of laws of conservation and equivalence. For the sake of convenience and simplicity, the events can be considered, which proceed at compression and expansion of the ideal gas in a cylinder with the plunger moving without friction. The theorem by R. Clausius is known for spontaneously proceeding process of expansion of the ideal gas, which is thermodynamically irreversible. This theorem is also applied to our proof in the first part of it.

5.4. Entropy of Thermodynamically Irreversible Spontaneous and Non-Spontaneous Processes

1. Compression and expansion of the ideal gas (expenses of energy for intermolecular interaction of molecules in the gas – attraction of molecules to each other during the gas expansion, and repulsion of molecules from each other during the gas compression – are absent) are really performed under the plunger of the cylinder operating irreversibly. But conditions can be also created for compression and expansion of the ideal gas, which enable the process to be quasi-equilibrium or equilibrium, as the border condition for the process conduction. These two types of processes – equilibrium and non-equilibrium –

differ by consumed and performed amounts of work, which will become clear from the proof of the theorem in below.

2. Expansion of gas. The process of expansion of the ideal gas, placed under the plunger, proceeds on account of energy addition to it in the heat form. This process can be illustrated by a graphic, depicted in Figure 5.2. For the sake of convenience, the process is conducted isothermally, T = const. At first, the process of gas expansion is conducted thermodynamically irreversible, from state 1 to state 2. For this purpose, pressure on the plunger is reduced initially down to point 3, and then the gas is allowed to change volume abruptly at P = const. Generally, the process proceeds by the broken line II, which is depicted by two straight lines for the sake of convenience. Evidently, this process proceeds continuously, and pressure and volume change continuously by a complicated curve, different from the curve II. The plunger produces work, which for thermodynamically irreversible process will be equal to $dF_1 = dW_{irrev.}$ – the square under the line 2-3 (or II).

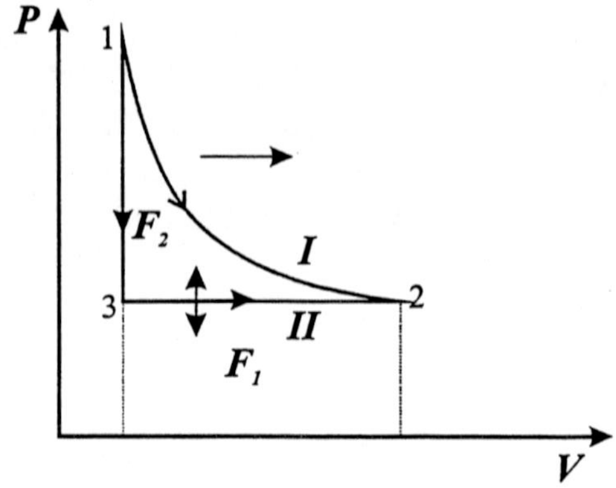

Figure 5.2. Illustration to the process of the gas expansion: I – reversible gas expansion; II – irreversible gas expansion at T = const.

Thermodynamically irreversible or, more accurate, quasi-equilibrium process of expansion of the ideal gas can be conducted between these two states of the ideal gas 1 and 2, characterized by values of parameters P_1, V_1 and P_2, V_2, i.e. the process can be conducted during infinitely long time, infinitely slowly, with the gas state infinitely close to the equilibrium, which is characterized by the curve I. In this case, the gas will conduct a work, determined as $dW_{reverse}$ – the square F_2 under the curve I. The relation of these works is determined according to appropriate squares under curves I and II.

Then, the work of the reversible process will be equal to the square

$$dF_2 = dW_{reverse}. \tag{5.1}$$

Work of the irreversible process will be equal to the square

$$dF_1 = dW_{irrev.}. \tag{5.2}$$

The inequality follows from comparison of squares ($dF_1 < dF_2$):

$$dW_{irrev.} < dW_{reverse}. \tag{5.3}$$

Transferring the work to the left part of the inequality, we will obtain difference of works smaller than zero:

$$dW_{irrev.} - dW_{reverse} < 0. \tag{5.4}$$

According to the law of equivalence, work is equal to heat. Then, instead of the difference of works, we obtain the inequality of heats:

$$dQ_{irrev.} - dQ_{reverse} < 0. \tag{5.5}$$

Dividing this inequality by temperature T, we get:

$$\frac{dQ_{irrev.}}{T} - \frac{dQ_{reverse}}{T} < 0. \tag{5.6}$$

The finite interval of change of the working substance state in the limits of the first and the second state, we integrate inequality (5.6):

$$\int_1^2 \frac{dQ_{irrev.}}{T} - \int_1^2 \frac{dQ_{reverse}}{T} < 0 \tag{5.7}$$

or

$$\int_1^2 \frac{dQ_{irrev.}}{T} < \int_1^2 \frac{dQ_{reverse}}{T}. \tag{5.8}$$

For thermodynamically reversible process, determined by the curve I, basing on the principle of existence of entropy, the integral from reduced heat equals to the difference of entropies:

$$\int_1^2 \frac{dQ_{reverse}}{T} = S_2 - S_1. \tag{5.9}$$

Substituting expression (5.9) into (5.8), we obtain the inequality for the difference of entropies in the following view:

$$S_2 - S_1 > \int_1^2 \frac{dQ_{irrev.}}{T}. \tag{5.10}$$

For adiabatic conditions, $dQ_{irrev.} = 0$, and then inequality (5.10) will determine growth of entropy for the process of spontaneous expansion of gas, which is characterized by inequalities:

$$S_2 - S_1 > 0 \text{ or } S_2 > S_1. \tag{5.11}$$

In the differential form, the change of entropy for a spontaneous process, proceeding in a phase-open system, will be greater than zero:

$$DS_{sp} > 0. \tag{5.12}$$

Differentiating dS by time, we obtain a regularity for entropy rise in the following view:

$$\left(\frac{\partial S_{sp}}{\partial \tau}\right)_T > 0. \tag{5.13}$$

This inequality is one of those for conjugated processes proceeding in phase-open systems. The second process in the considered example is non-spontaneous compression of gas. The process proceeds at expense of work produced by external forces.

Compression of the Ideal Gas – Non-Spontaneously Proceeding Process

The process of gas compression can be conducted differently – by loading of heavy loads of mass m_1, influence of compressed coil spring on a plunger, etc. Compression of gas can proceed at a finite rate, i.e. is thermodynamically irreversible. This process is shown by curve II in Figure 5.3, which is simply presented by an isochore and isobar. Evidently, the curve of compression in the irreversible process is of a complex form, but this does not change the sense of the proof.

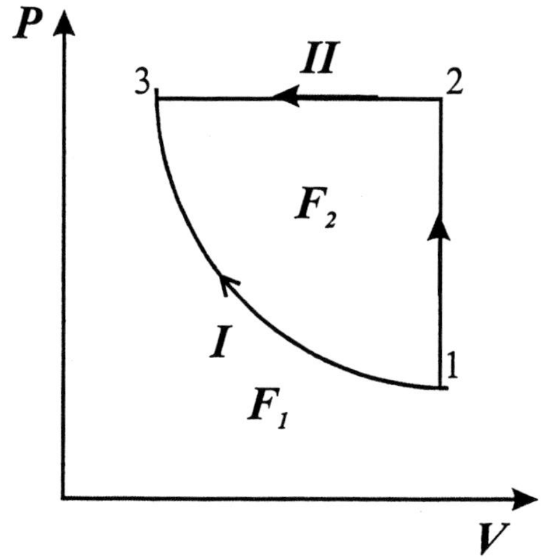

Figure 5.3. Compression of the ideal gas.

In the present case, compression of gas can be conducted quasi-statically by the curve I. Squares under curves II and I determine works consumed for irreversible and quasi-equilibrium compression of gas. These works are determined by equations:

$$dF_2 = dW_{irrev.};\ dF_1 = dW_{reverse}. \tag{5.14}$$

The difference in squares will be greater than zero:

$$dF_2 - dF_1 > 0. \tag{5.15}$$

Substituting (5.14) into (5.15), we obtain the inequality for works consumed for gas compression in irreversible and reversible processes:

$$dW_{irrev.} - dW_{reverse} > 0. \tag{5.16}$$

Consequently, the difference of works in this case will be grater than zero. Taking into account the law of equivalence, works can be substituted by heats, and the following inequality for the difference of heats can be obtained:

$$dQ_{irrev.} - dQ_{reverse} > 0. \tag{5.17}$$

Dividing inequality (5.17) by the absolute temperature T, we get:

$$\frac{dQ_{irrev.}}{T} - \frac{dQ_{reverse}}{T} > 0. \tag{5.18}$$

In this inequality, reduced heats are presented for irreversible and equilibrium processes. For equilibrium processes, the principle of existence of entropy in the following analytical form is true:

$$\frac{dQ_{reverse}}{T} = dS. \tag{5.19}$$

Substituting (5.19) into (5.18), we obtain that the reduced heat in the irreversible process is higher than changes of entropy in the reversible process at the sacrifice of various *noise effects* in the working substance (local overheating, turbulization and other processes). Consequently, the inequality can be presented in the following view:

$$\frac{dQ_{irrev.}}{T} > dS. \tag{5.20}$$

For the sake of convenience of analysis of the processes, an adiabatic system can be used, for which heat exchange with the surrounding is absent. Then

$$dQ_{irrev.} = 0. \tag{5.21}$$

Substituting (5.21) into (5.20), we obtain

$$dS_{n\text{-}sp} < 0, \tag{5.22}$$

i.e. conduction of a spontaneous process (cooling down of the working substance below temperature of the surrounding, compression, concentration of solutions, condensation, etc.) leads to a decrease of entropy of the working substance. Difference in entropies with the minus sign is called neg-entropy. The regularity of growth of neg-entropy, i.e. decrease of entropy, in the system is determined by the inequality

$$\left(\frac{\partial S_{n-sp}}{\partial \tau}\right)_T < 0. \tag{5.23}$$

5.5. Thermodynamic Proof of the Ratio of Inequalities for Spontaneous and Non-Spontaneous Processes

Direction of proceeding of processes and change of entropy in phase-open systems and the surrounding can be proved considering work of a thermal machine under non-stationary conditions. Thermal machine includes a heater, a device for work production at the sacrifice of heat redistribution between the heater and the cooler, and a cooler. Structure of the machine is shown in Figure 5.4.

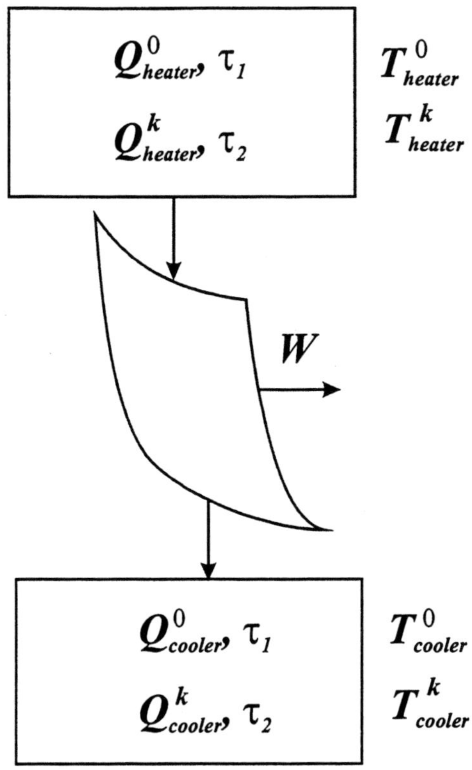

Figure 5.4. Operation of the thermal machine under non-stationary conditions, τ is time.

The machine operates at the sacrifice of detachment of some amount of heat from the heater under non-stationary, transformation of a part of heat into work and transmission of the rest of heat to the cooler.

Let up to time τ_1 heat of the heater, transmitted to the working substance of the machine at temperature T_1^0, be equal to Q_h^0; up to time τ_2 heat of the heater will decrease to Q_h^k at T_1^k. In this case, according to the supposition, the value $Q_h^0 > Q_h^k$ in the absence of additional source, compensating heat decrease, where *0* and *k* are intermediate states of the heater by times τ_1 and τ_2.

Difference of heats will be smaller than zero:

$$Q_h^k - Q_h^0 < 0. \tag{5.24}$$

Dividing these heats by T_1^k and T_1^0, respectively, we obtain:

$$\frac{Q_h^k}{T_1^k} - \frac{Q_h^0}{T_1^0} < 0. \tag{5.25}$$

For short time interval it can be assumed that temperatures are low different from each other, i.e. $T_1^k \cong T_1^0$. Then the following value of the reduced heat can be easily obtained:

$$-\frac{\Delta Q_h}{T_h} < 0. \tag{5.26}$$

If the time interval is decreased, i.e. under the condition $\Delta\tau \to 0$, we obtain

$$-\frac{dQ_h}{T_h} < 0. \tag{5.27}$$

Differentiation of this expression of the reduced heat by time gives:

$$\frac{dQ_h}{d\tau} \times \frac{1}{T_h} > 0. \tag{5.28}$$

The reduced heat equals to entropy, that is why inequality (5.28) can be easily reduced to the following view:

$$\frac{dS_h}{d\tau} > 0. \tag{5.29}$$

This inequality displays that for energy spontaneously released by the system (heater) in the heat form under non-stationary conditions, i.e. for the process of heat transmission from the system to the working substance, entropy growth. For the heat flow, entropy grows at energy yielding the flow element.

Similar discussion can be used for distribution of heat in the cooler. At the initial moment of time τ_1 the amount of heat Q_{cooler}^0 is added to the cooler, and up to the moment τ_2 the amount of heat added to the cooler will equal to Q_{cooler}^k. The amounts of heat absorbed by the cooler grow with time, so the inequality is true:

$$Q_{cooler}^k > Q_{cooler}^0. \tag{5.30}$$

The difference will be greater than zero:

$$Q_{cooler}^k - Q_{cooler}^0 > 0. \tag{5.31}$$

Changing signs, we obtain:

$$-\left(Q^k_{cooler} - Q^0_{cooler}\right) < 0 \text{ or } Q^0_{cooler} - Q^k_{cooler} < 0. \tag{5.32}$$

Dividing this by T^0_{cooler} and T^k_{cooler}, we obtain:

$$\frac{Q^0_{cooler}}{T^0_{cooler}} - \frac{Q^k_{cooler}}{T^k_{cooler}} < 0. \tag{5.33}$$

For the condition $\Delta\tau \to 0$, we may assume the equation $T^k_{cooler} \cong T^0_{cooler}$ and obtain the following inequalities:

$$\frac{\Delta Q_{cooler}}{T_{cooler}} < 0 \text{ or } \frac{dQ_{cooler}}{T_{cooler}} < 0. \tag{5.34}$$

Differentiating these expressions by τ, we obtain

$$\frac{dQ_{cooler}}{T_{cooler} d\tau} < 0 \text{ or } \left(\frac{\partial S_{cooler}}{\partial \tau}\right)_{x_i} < 0. \tag{5.35}$$

Consequently, a non-spontaneous process of heat transmission from the working substance to the cooler is accompanied by reduction of entropy.

Heat transmission from the heater and its absorption by the working substance is the spontaneous process (heating up of the working substance) and is accompanied by growth of entropy of the working substance. Heat elimination by the cooler from the working substance is the non-spontaneous processes and is accompanied by reduction of entropy of the working substance.

5.6. Changing of Entropy in a Phase-Open System Interacting with the Surrounding

In phase-open systems interacting with the surrounding (or exchanging energy and substance between internal parts of the system), spontaneous and non-spontaneous processes proceed continuously. Spontaneous processes proceeding in the range of a single system are decelerated by non-spontaneous ones. At the same time, spontaneous process of one phase-open system is the source of proceeding of a non-spontaneous process in another system. For example, energy radiation by one system causes absorption of this energy in another system, what leads to heating up of it, pressure boost in it and change of other states of the system.

In the first system, store of the free energy decreases with temperature, and entropy of the system, whereas store of the internal energy in the second system, interacting with conjugation with the first one, grows, and entropy of the second system growing.

Consequently, total change of entropy – for phase-open systems interacting with the surrounding – must be determined by the equation, which should include derivatives of entropy change by time for spontaneous and non-spontaneous processes. This is determined by the following equation:

$$\left(\frac{\partial S_{sp}}{\partial \tau}\right)_{x_i} + \left(\frac{\partial S_{n-sp}}{\partial \tau}\right)_{x_i} = \left(\frac{\partial \Delta S_{res}}{\partial \tau}\right), \tag{5.36}$$

$$\left(\frac{\partial \Delta S_{res}}{\partial \tau}\right)_{x} = \Delta,$$

where *res* is the resulting value.

The expression can be considered as an analytical form of the fourth law of thermodynamics. This law could not appear from consideration of equilibrium systems and processes only. Its formulation has appeared as a result of broadening fundamental grounds of thermodynamics of irreversible processes, in presentation by N.I. Belokon' and I.R. Prigozhin – as broadening of R. Clausius theorem.

The value of parameter Δ can be greater than zero, equal to zero and smaller than zero. The value $\Delta > 0$ means prevailing of a spontaneous process over a non-spontaneous one. $\Delta < 0$ means prevalence of action of a non-spontaneous process over a spontaneous one. The value $\Delta = 0$ means complete compensation of a spontaneous process by a non-spontaneous one, and vice versa. This is typical for equilibrium processes.

One can try to show examples of the above-mentioned types of processes, characterized by the parameter Δ. The number of similar examples is infinitely great, but they can be easily selected if necessary.

As the first example, when $\Delta > 0$, the process of transition over-saturated vapor into liquid with formation on saturated vapor. The second example, when $\Delta < 0$, is the process of formation of over-saturated vapor from saturated one. The first process can be the source of work, the second one proceeds only at the work expense or addition of energy to the system.

The important consequences of the fourth law of thermodynamics are the following:

- According to this law, it is indifferent to the matter, where and during what time formation of its material state is possible, or vice versa – transition of its material state into the absolute vacuum. Material and other states of the matter are continuously formed and disappear in the surrounding at the sacrifice continuous, interconnected and conjugated proceeding of spontaneous and non-spontaneous processes;

- The idea about the beginning of development of the Universe from grounds of the fourth law of thermodynamics becomes senseless. Such beginning could be presented as existence of the only one of processes separately from another in nature, i.e. spontaneous separately from non-spontaneous one only, and vice versa;
- The idea about the end of the World of the Universe is also senseless, because according to expression (5.24) and consequences, yielding from it, the value of Δ is finite for interactions of phase-open systems and the surrounding only. Compression of the World or the Universe excludes such interaction;
- New thermodynamic correlations must consider different signs of the change of Δ, and these signs must be calculated in general equations of thermodynamics of irreversible processes.

Chapter 6

Correlation of Processes for Interacting Phase-Open Systems and the Surrounding

6.1. Direction of the Process Proceeding

1. Phase-open systems continuously interact with the surrounding, exchanging substance and energy. In this case, the following main directions of such exchange can be selected.

A. A phase-open system *transmits* to the surrounding:

- Excess of the substance at constant temperature *T*, pressure *P* and volume *V*;
- Energy in the heat or another form at constant mass *m*, pressure *P* and volume *V*;
- Energy and substance.

B. A phase-open system *accepts* from the surrounding:

- Substance at constant pressure *P*, temperature *T* and volume *V*;
- Energy at constant pressure *P*, temperature *T* and volume *V*;
- Energy and substance.

In the first case (A), the surrounding plays the role a decelerator for processes proceeding in phase-open systems as a result of their interaction with each other.

In the second case (B), deceleration is performed by a phase-open system, in which contractions to processes proceeding in the surrounding may appear.

2. Numerous examples can be displayed, which enable the above-mentioned directions of preferable redistribution of substance and energy to be presented more clearly under interaction of the system and the surrounding. In winter time, apartments heated up from inside spontaneously transmit heat to the surrounding without transmission of the substance, and in summer time apartments are heated up by energy, added to them by the surrounding.

Open basins exchange substance and energy with the surrounding, and direction of the process depends on external parameters of the surrounding and internal parameters of basins. In summer time, evaporation of basins (non-spontaneous process) prevails over condensation of saturated water vapor. In wintertime, reverse processes prevail. Conditions of prevalence of spontaneous processes over non-spontaneous ones, and vice versa, for biological objects can be selected. However, this is a separate problem; in this case, the most important problem is origin, development and degeneration of biological objects.

6.2. The Example of Heat Processes with Diminution and Growth of Entropy

1. An open system can be homogeneous by composition and structure or include several subsystems, differed by temperature, pressure and composition, content of internal and free energy. Let a simple system be in contact with the surrounding and able to exchange substance and energy with it, as shown in Figure 6.1. This is also true for complex systems, but analytical equations for simple systems is more simple, too.

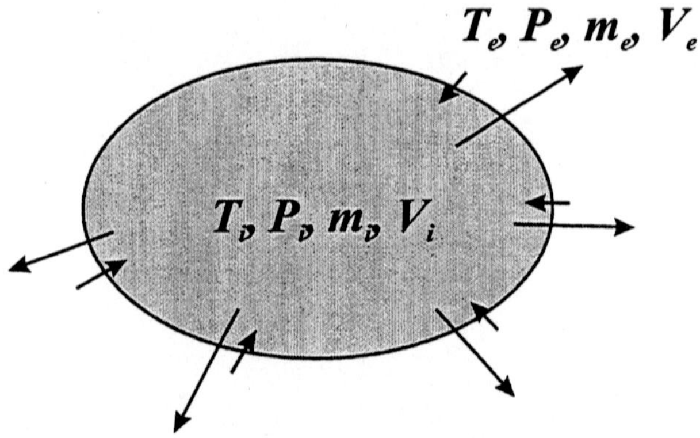

Figure 6.1. The scheme of the phase-open system interaction with the surrounding.

Considering interaction of the phase-open system with the surrounding, two real cases can be separated:

1) Intensive parameters of working substances in the system are greater than those in the surrounding;
2) Intensive parameters of the surrounding are greater than those of working substances in the phase-open system.

In the first case, transmission of substance and energy in different forms of it from the phase-open system to the surrounding happens. This process proceeds up to leveling of intensive parameters in the system with these of the surrounding. In this case, it is assumed that the phase-open system and the surrounding transit into the equilibrium. But in the real medium, various continuous and interconnected fluctuation micro- and macroprocesses proceed, which enable phase-open systems and the surrounding not to exist in the state of equilibrium. Homogeneous values of thermodynamic functions and, as a consequence, intensive parameters, can appear just for a short moment. Consequently, the idea of equilibrium in real media is the border notion for the state of the real system and the surrounding, which cannot be obtained for long in the real systems.

When energy in the heat form is transmitted from a homogeneous phase-open system to the surrounding, temperature in the phase-open system decreases from T_{ini} to $T_{surround}$ (from initially heated state of the substance to temperature of the surrounding), moreover, $T_{ini} > T_{surround}$. In this case, the phase-open system is cooled down, and entropy is decreased, which corresponds to the following thermodynamic equation:

$$\Delta S_I^{sys} = C_P \ln \frac{T_{surround}}{T_{ini}} < 0, \; \Delta S_I^{surround} = C_P \ln \frac{T_{surround}}{T_{ini}} > 0, \quad (6.1)$$

where C_P is the heat capacity of the working substance (or substances) in the phase-open systems.

In this case, entropy of the system decreases, and entropy of the surrounding grows in the range of a local interaction of the phase-open system with the surrounding.

2. In the second case, intensive parameters in the phase-open system are smaller by value than these of the surrounding. In such conjugated systems interacting with the surrounding, redistribution of energy and substance proceeds with preferable flow of them to the side of the phase-open system, as shown in Figure 6.2.

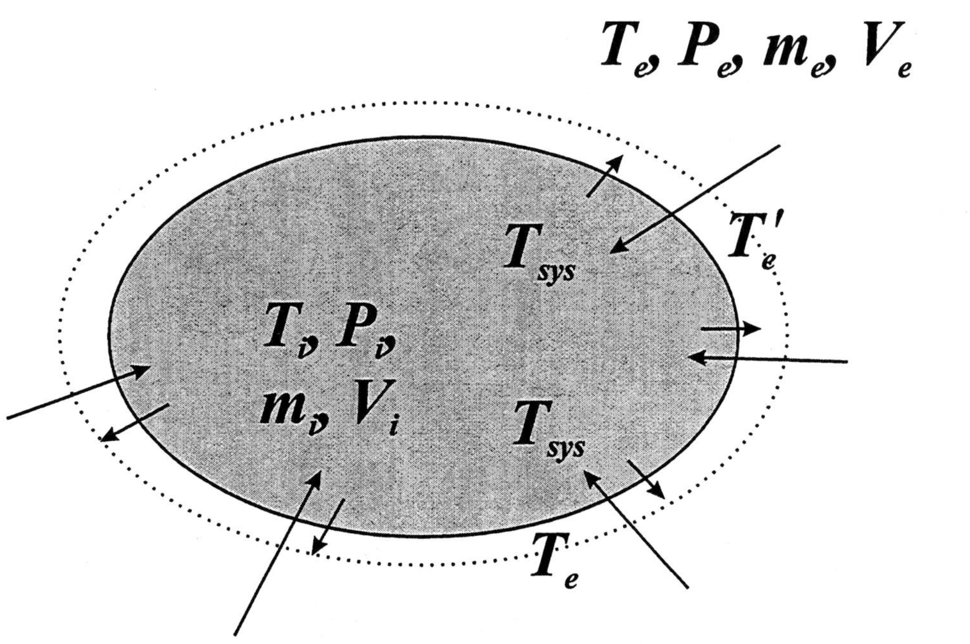

Figure 6.2. Scheme of interaction of the phase-open system with the surrounding.

When energy and substance are transmitted into the phase-open system, values of internal temperature T_i, pressure P_i and chemical potentials μ_i (energy per mole of transmitted substance under constancy of natural variables) grow in it. These assumptions are conditional, because in irreversible processes it can be said only about immediately stable state of the system.

Under such interaction with the phase-system, temperature of the surrounding T_e decreases, and temperature of the phase-open system rises. Therewith entropy of a phase-open system grows with time according to the equation:

$$\Delta S_2 = C_P \ln\frac{T_e}{T_{sys}} > 0, \tag{6.2}$$

where T_{sys} is temperature of the phase-open system ($T_{sys} < T_e$), T_e is temperature of the surrounding. In this case, entropy in the phase-open system grows, and decreases in the surrounding with time, according to the equation

$$\Delta S_2' = C_P \ln\frac{T_{sys}'}{T_e} < 0, \tag{6.3}$$

where T_{sys}' is temperature of a part of the space neighbor to the phase-open system; T_e is temperature of the surrounding ($T_{sys}' < T_e$).

3. These equations reflect no change of entropy with time. It must be noted that in the first case spontaneous process of transmission of free energy and entropy to the surrounding prevails. In this case, sources of entropy and energy are limited. Energy and entropy were accumulated by the system during previous processes of its interaction with the surrounding.

In the case of leveling of intensive parameters in the surrounding, changes in the system can proceed under the effect non-spontaneous processes only, proceeding in the surrounding (radiation and heat energy, work, etc.). In the second case, non-spontaneous processes can prevail over spontaneous ones in the phase-open systems.

6.3. Interaction of System, Including Subsystems, with the Surrounding

1. A heterogeneous system can include a great number of subsystems, characterized as phase-open parts of the system. These phase-open subsystems exchange energy, substance and entropy with each other and can be also characterize by a definite store of free and bound energy. Structure of the system with subsystems is depicted in Figure 6.3.

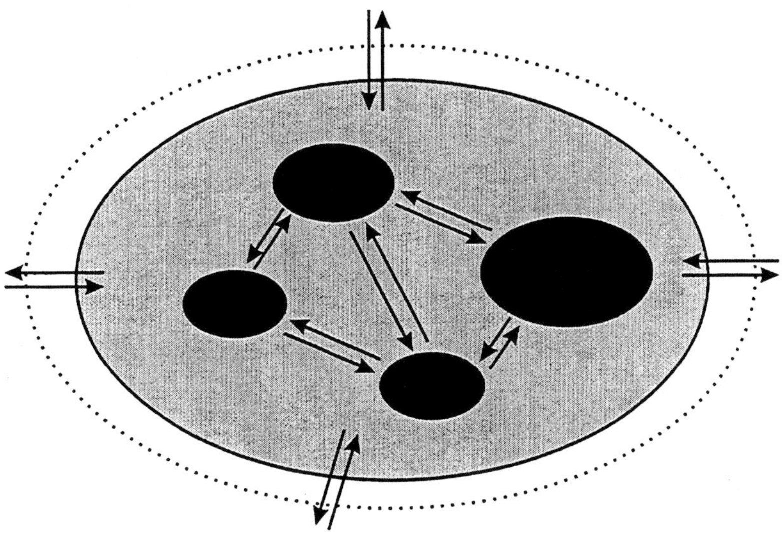

Figure 6.3. Illustration to the system with subsystems.

Energy, substance and entropy are redistributed between subsystems in accordance with values of intensive parameters (P, T, μ_i, and others). The process will proceed until intensive parameters of all subsystems are leveled, and the system levels intensive parameters with the surrounding.

2. Total values of heats, internal energies, entropies and works for the system will equal to:

$$d_iQ = \Sigma d_iQ_i,\ d_iU = \Sigma d_iU_i,\ d_iW = \Sigma d_iW_i,\ d_iS = \Sigma d_iS_i. \tag{6.4}$$

Analytical expression for heat of the system will be of the following view:

$$d_iQ = \Sigma d_iQ_i = \Sigma d_iU_i + d_iW = \Sigma T_i d_iS_i. \tag{6.5}$$

The surrounding in the range of the order closest to the system will possess the following values of heat, internal energy, work and entropy:

$$d_eQ = \Sigma d_eQ_i,\ d_eU = \Sigma d_eU_i,\ d_eW = \Sigma d_eW_i,\ d_eS = \Sigma d_eS_i. \tag{6.6}$$

General expression of heat will include all these values:

$$d_eQ_i = \Sigma d_eU_i + d_eW_i = \Sigma T_e d_e S_i. \tag{6.7}$$

Total equation can be presented in the following form:

$$dQ = d_iQ + d_eQ = \Sigma d_iQ_i + \Sigma d_eQ_i. \tag{6.8}$$

Substituting (6.5) and (6.7) into (6.8), we obtain general expression for interaction of a complex phase-open system with the surrounding:

$$dQ = \Sigma d_eU_i + \Sigma d_iU_i + \Sigma d_eW_i + \Sigma d_iW_i = \Sigma T_e d_e S_e + \Sigma T_i d_i S_i. \tag{6.9}$$

To determine the type of the process, equation (6.9) is integrated in the range from the first to the second and from the second to the first state.

3. For spontaneous processes, total values *dQ* will be positive, and for non-spontaneous ones – negative, i.e.

$$dQ \overset{>}{\underset{<}{=}} 0, \tag{6.10}$$

$$TdS = \sum T_e d_e S_i + \sum T_i d_i S_i dQ \overset{>}{\underset{<}{}} 0. \tag{6.11}$$

Consequently, interaction of phase-open systems with the surrounding is characterized by different signs, reflecting direction of excessive processes, which lead to leveling of intensive parameters of the system and the surrounding. The sign *minus* determines transmission of energy and substance from the system to the surrounding, the sign *plus* – transmission of energy and substance from the surrounding to the system; opposite signs are used for changes of entropy.

The sign *minus* for change of entropy marks its decrease in the system, and the sign *plus* – growth of entropy in the surrounding.

Chapter 7

KINETICS OF ENTROPY VARIATION

7.1. MAIN CONCEPTS

1. Phase-open systems and the surrounding permanently exchange substance and energy, which permanently change value of the free energy in the system and the surrounding.

A phase-open system consisted of k-subsystems displays redistribution of substance and energy between these subsystems, and simultaneously subsystems exchange energy and substance with the surrounding. This redistribution of energy and substance bound to entropy change can be presented by the following expression:

$$\frac{1}{T}\cdot\frac{dQ}{d\tau}=\frac{1}{d\tau}\left[\left(\frac{d_iQ_1}{T_1^i}-\frac{d_iQ_2}{T_2^i}\right)+\left(\frac{d_iQ_2}{T_2^i}-\frac{d_iQ_3}{T_3^i}\right)+\ldots+\left(\frac{d_iQ_k}{T_k^i}-\frac{d_iQ_{k+1}}{T_{k+1}^i}\right)\right]+$$
$$+\frac{1}{d\tau}\left[\left(\frac{d_iQ_1'}{T_1^i}-\frac{d_eQ_1}{T_1^e}\right)+\left(\frac{d_iQ_2'}{T_2^i}-\frac{d_eQ_2}{T_2^e}\right)+\ldots+\left(\frac{d_iQ_k'}{T_k^i}-\frac{d_eQ_k}{T_k^e}\right)\right]+ \tag{7.1}$$
$$+\frac{dW_i}{d\tau}+\sum\frac{\mu^i d_i n_i}{d\tau}+\sum\frac{\mu^e d_e n_e}{d\tau}+\sum\frac{\mu_j^{ie} d_e^i n_i^e}{d\tau},$$

where the sum of chemical potentials reflects physicochemical processes inside the system, in the surrounding, and mass transfer of substances between the surrounding and the system; $\frac{dW_i}{d\tau}$ is any work; ${}_i\mu_i$ are chemical potentials of substances in the system; ${}_e\mu_i$ are chemical potentials of substance in the surrounding; μ_i^{ie} are chemical potentials at the interphase with the surrounding; ${}_iQ_i, {}_eQ_i'$ are heats transmitted between subsystems and the surrounding.

2. Taking into consideration that reduced heat equals to entropy, equation (7.1) may be presented in the following form:

$$\frac{dS}{d\tau}=\frac{d_iS_1}{d\tau}-\frac{d_iS_2}{d\tau}+\frac{d_iS_2}{d\tau}-\frac{d_iS_3}{d\tau}+\ldots+\frac{d_iS_k}{d\tau}-\frac{d_iS_{k+1}}{d\tau}+\dot{W}_i+$$
$$+\sum\frac{d_iS_i'}{d\tau}-\sum\frac{d_eS_i'}{d\tau}+\sum\frac{\mu^i d_i n_i}{d\tau}+\sum\frac{\mu^e d_e n_i}{d\tau}+\sum\frac{\mu_j^{ie} d_e^i n_i^e}{d\tau}. \tag{7.2}$$

This equation determines entropy flow inside the system and exchange of entropy between the system and the surrounding, as well as change of entropy due to exchange of substance between parts of the system and with the surrounding. The resulting flow can be directed both from the system into the surrounding and vice versa.

In the first case, influence of the surrounding on the system state is more than overcome by the system and, hence, the main role in the resulting process is played by spontaneous processes, proceeding in the system (for example, radioactive decay of isotopes, heat transmission from more heated system into the surrounding, etc.). In this case, the positive sign can be prescribed to the entropy flow into the surrounding. Heat flow inside the system decreases. Heat transmission from the system to the surrounding is determined as an exothermal process, and the sign minus is prescribed to this flow.

In the closed system entropy grows and approximates the maximal value, because the source of the resulting entropy flow strives to zero. In the open system entropy inside the system will decrease at the sacrifice of cooling down the working substance, and entropy of the surrounding will increase due to heating up of it.

3. Under definite conditions, state of the system can exist at lower values of intensive parameters comparing with these of the surrounding. Under such state of the system and the surrounding, the resulting flow of energy and entropy will be directed from the surrounding to the system. The system undergoes the influence of spontaneously proceeding processes (expense of energy or work, substance transmission from the surrounding to the system, and other interactions). In this case, entropy and free energy excess will grow. In the surrounding, vice versa, entropy and free energy will decrease. The resulting flow of entropy and energy will be directed from the surrounding to the system. Heat transmission from the surrounding to the system is the endothermal process, and the positive sign is ascribed to it. A non-spontaneous is characterized by growth of the free energy store (ΔA, ΔG), and the positive sign is ascribed to it, too.

7.2. Entropy of Substance Flow in the Open System

7.2.1. General Concepts

1. Equation of entropy flow of substances for open systems can be deduced on the basis of equations of the energy flow in the heat form for open systems also. Flows of substances, energy and entropy for processes proceeding in nature and industry are shaped as air masses (winds, hurricanes, etc.), springs and rivers, streams (Gulf Stream), smoke pipe outflows, flows of gases and liquids in pipelines, reactors, etc. Energy and substance of flows permanently exchange excess of substances and energy, as well as various types of work with the surrounding.

Physical, mechanical, chemical and biological processes proceed in flows of substances. Each of the processes can be presented by equations including the number of terms, which reflects the main mechanism of the energy (entropy) and substance transmission by coordinates of the flow.

2. Physical processes include electromagnetic and electrostatic, Van der Waals interactions, processes of the phase transition of substances, etc.

Transmission of energy from one part of the system (flow) to another may include the following mechanisms: molecular, convection, the one bound to heat conductivity, heat transmission, radiation, and other types.

Hydromechanical flow is also characterized by transfer of substance by convection, pressure drop, difference in transfer rates of mass of substances, etc.

3. First of all, consider behavior and state of the flow from the energetic point of view. Flow of energy in the heat form is induced by convectional forces, heat conductivity, heat transmission, chemical transformation of substance, consumption of different types of works. This flow may also exist under non-stationary or stationary conditions.

Applying the general law of conservation to the flow of substances under non-stationary conditions and taking into account the mechanism of energy transmission in the form of heat from one part to another, total equation of the heat flow can be deduced. Let first deduce equations of flows for its separate components for non-stationary conditions.

7.2.2. Equation of Convectional Heat Flow

Total amount of heat, injected into the flow of elementary volume dV during time $d\tau$, will be determined by the equation:

$$dQ_{inj} = C_p \rho \upsilon dV d\tau, \tag{7.3}$$

where C_p, ρ, υ, V, and τ are heat capacity, density, linear velocity, volume and time, respectively.

Heat transmission under non-stationary conditions in the range of assigned elementary volume of the flow is accompanied by decrease or increase of yielding heat comparing with its amount entering the flow, which is determined by equation of the following view:

$$dQ_{yield} = C_p \rho \upsilon dV d\tau + \frac{\partial T}{\partial x} C_p \rho \upsilon dV d\tau. \tag{7.4}$$

Accumulation of energy in the elementary volume dV during time $d\tau$ may be determined, subtracting expression (7.3) from equation (7.4):

$$dQ_{yield} - dQ_{inj} = \frac{\partial T}{\partial x} C_p \rho \upsilon dV d\tau . \qquad (7.5)$$

According to the law of conservation, the amount of energy accumulated in a flow of assigned configuration must be equal to the amount of energy, yielded from the assigned volume:

$$dQ_{cons} = -\frac{\partial(\rho C_p T)}{\partial \tau} dV d\tau , \qquad (7.6)$$

where Q_{cons} is the quantity of heat at the exit of the flow, kJ/m^{3}s.

Equating equations (7.5) and (7.6), we obtain an expression for calculation of convective force of the flow, moving along the coordinate x, as follows:

$$\upsilon \rho C_p \frac{\partial T}{\partial x} = -\frac{\partial(\rho C_p T)}{\partial \tau} . \qquad (7.7)$$

In the heat form, energy in the flow is transmitted by convection in directions of coordinates x, y and z, simultaneously. Then for conditions of non-stationary state of the flow, the equation can be presented in the form of the flow gradient $C_p\rho\upsilon\,\mathrm{grad}T$ or in the following extend view:

$$\rho C_p \upsilon_x \frac{\partial T}{\partial x} + \rho C_p \upsilon_y \frac{\partial T}{\partial y} + \rho C_p \upsilon_z \frac{\partial T}{\partial z} = -\frac{\partial(\rho C_p T)}{\partial \tau} . \qquad (7.8)$$

This equation does not reflect the aggregate state of the substance, in the frames of which energy in the heat form is transferred, and nature of the substance is reflected by values of density and heat capacity of the substance.

7.2.3. Heat Transfer by Heat Conductivity

Heat conductivity is bound to energy transfer via collision of molecules with each other, change of oscillation and rotational energy of molecules, and to other forms of energy transmission.

Heat conductivity determines spreading of heat in the three-dimensional system at inequality of temperatures at different removals of heat flow from the initial point. Differential equation of heat conductivity may be deduced, if transmission of energy in the heat form in the frames of cube with x, y and z edges is considered.

The Fourier law is taken into account in deduction of the heat conductivity equation. This law determines that the amount of heat dQ, transmitted by heat conductivity, is proportional to the temperature gradient dT/dx, time of change of the flow state and cross-section of the flow dF, perpendicular to the flow:

$$dQ_x = -\lambda \frac{dT}{dx} dF d\tau , \tag{7.9}$$

where λ is the heat conductivity coefficient.

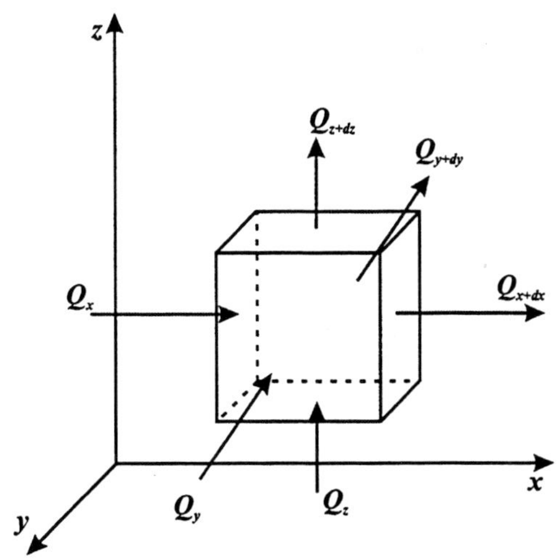

Figure 7.1. Scheme of the energy flow in the heat form.

From this equation, heat conductivity coefficient determines the amount of energy, which due to heat conductivity passes through the cross-section of 1 m^2 during specific time at 1 degree difference in temperatures per 1 m of the normal to the isothermal surface. In the general case, the heat conductivity coefficient depends on T, P and nature of the substance.

Through separate sides Q_x, Q_y and Q_z of heat enter the cube during time $d\tau$, and the following amounts of heat leave it from the opposite sides: Q_{x+dx}, Q_{y+dy} and Q_{z+dz}. Accumulation of heat in the assigned volume will be determined by the expression:

$$dQ_{accum} = (Q_x - Q_{x+dx}) + (Q_y - Q_{y+dy}) + (Q_z - Q_{z+dz}). \tag{7.10}$$

According to the Fourier law, heat conductivity injects heat into the cube in amount, determined by heat expressions:

$$Q_x = -\lambda \frac{\partial T}{\partial x} dydzd\tau, Q_y = -\lambda \frac{\partial T}{\partial y} dxdzd\tau, Q_z = -\lambda \frac{\partial T}{\partial z} dxdyd\tau . \tag{7.11}$$

Cross-sections of the cube will yield heat in amount, determined by formulae:

$$Q_{x+dx} = -\lambda\frac{\partial T}{\partial x}dydzd\tau - \lambda\frac{\partial\left(\frac{\partial T}{\partial x}dx\right)}{\partial x}dydzd\tau = -\lambda\frac{\partial T}{\partial x}dydzd\tau - \lambda\frac{\partial^2 T}{\partial x^2}dxdydzd\tau, \tag{7.12}$$

$$Q_{y+dy} = -\lambda\frac{\partial T}{\partial y}dxdzd\tau - \lambda\frac{\partial^2 T}{\partial y^2}dxdydzd\tau, \tag{7.13}$$

$$Q_{z+dz} = -\lambda\frac{\partial T}{\partial z}dxdyd\tau - \lambda\frac{\partial^2 T}{\partial z^2}dxdydzd\tau. \tag{7.14}$$

Subtracting expressions (7.12)-(7.14) from (7.11), we obtain heat accumulation in the cube in the form of the following expression:

$$dQ_{accum} = \lambda\left(\frac{\partial^2 T}{\partial x^2} + \frac{\partial^2 T}{\partial y^2} + \frac{\partial^2 T}{\partial z^2}\right)dxdydzd\tau. \tag{7.15}$$

In accordance with the law of conservation, accumulation equals to consumption in the heat flow:

$$dQ_{accum} = dQ_{cons}, \tag{7.16}$$

where

$$dQ_{cons} = C_p\rho dxdydz\frac{\partial T}{\partial \tau}d\tau. \tag{7.17}$$

Substituting (7.15) and (7.17) into (7.16), we obtain differential Fourier equation for heat conductivity:

$$\lambda\left(\frac{\partial^2 T}{\partial x^2} + \frac{\partial^2 T}{\partial y^2} + \frac{\partial^2 T}{\partial z^2}\right) = \frac{\partial T}{\partial \tau}C_p\rho. \tag{7.18}$$

To simplify the equation, temperature conductivity coefficient is introduced in it:

$$a = \frac{\lambda}{C_p\rho}. \tag{7.19}$$

Then the Fourier formula is transformed into the following view:

$$a\left(\frac{\partial^2 T}{\partial x^2}+\frac{\partial^2 T}{\partial y^2}+\frac{\partial^2 T}{\partial z^2}\right)=\frac{\partial T}{\partial \tau}. \qquad (7.20)$$

7.2.4. Heat Transmission by Heat Emission Through the Interphase

The first Fourier law for heat transmission through the wall can be transformed for a linear law, changing differentials via finite differences, i.e.

$$Q_{heatf}=-\lambda\frac{\Delta T}{\Delta x}dydzd\tau. \qquad (7.21)$$

Or in another form:

$$Q_{heatf}=\alpha\left(\frac{T_{wall}-T_{surr}}{x_2-x_1}\right)dFd\tau, \qquad (7.22)$$

where Q_{heatf} is the heat flow; T_{wall} and T_{surr} are temperatures of the wall and the surrounding, respectively.

Then we introduce designation of the heat transmission coefficient:

$$\alpha=\left[\frac{dQ}{(T_{wall}-T_{surr})dFd\tau}\right]\frac{J}{cm^2grad}=\frac{W}{m^2grad}. \qquad (7.23)$$

As a consequence, we obtain the following equation of heat emission:

$$dQ_{heatf}=\alpha(T_{wall}-T_{surr})dFd\tau. \qquad (7.24)$$

Consequently, the amount of heat dQ_{heatf}, transmitted from the heat exchange surface to the surrounding (inside the flow) or outside it, or vice versa, is proportional to the heat exchange surface dF, difference in temperatures of the surface T_{wall} and the surrounding T_{surr}, and heat transmission time $d\tau$.

7.2.5. Production of Heat by Chemical Reaction

A complex of chemical reactions between organic and inorganic molecules may proceed in the flow of substances. During these reactions energy is absorbed and excreted in the heat form, the total process being endothermic, when the reactions proceeding in the flow absorb greater amount of heat, than excrete, and vice versa. In the first case, heat is delivered from the surrounding, and in the second case, excess of energy in the heat form is excreted from the system into the surrounding. Change of state of the flow of substances due to heat of a chemical process under non-stationary conditions may be determined by the equation of the heat balance of the following view:

$$\sum_{i}^{n} \frac{n_A^0 dx}{Fdl} \Delta H_{T,i} dV d\tau = \frac{-\partial(\rho C_p T)}{\partial \tau} dV d\tau, \tag{7.25}$$

where $\Delta H_{T,i}$ is the heat of the chemical process, calculated by the Hess law for constant temperature:

$$\Delta H_{T_i} = \sum \nu_i' \Delta H_{A_i'} - \sum \nu_i \Delta H_{A_i}, \tag{7.26}$$

and by the Kirchhoff formula for variating temperature:

$$\frac{d\Delta H_{T_i}}{dT} = \sum \nu_i' \Delta H_{A_i'} - \sum \nu_i \Delta H_{A_i}, \tag{7.27}$$

where ν_i' and ν_i are stoichiometric coefficients before indexes of reaction products and initial substances, respectively; $\Delta H_{A_i'}$ and ΔH_{A_i} are changes of enthalpies of reaction products and initial substances, respectively.

7.3. General Equation of the Heat Flow

General equation of the heat flow, without external heat exchange any kind, includes corresponding equations of the flow, displayed in the previous Section, and can be written down in the following form:

$$\begin{gathered} \rho C_p \upsilon_x \frac{\partial T}{\partial x} + \rho C_p \upsilon_y \frac{\partial T}{\partial y} + \rho C_p \upsilon_y \frac{\partial T}{\partial y} - \lambda\left(\frac{\partial^2 T}{\partial x^2} + \frac{\partial^2 T}{\partial y^2} + \frac{\partial^2 T}{\partial z^2}\right) + \\ + \alpha(T_{wall} - T_{surr}) + \sum_{i}^{n} w_{P_i} \Delta H_{H_i} = -\frac{\partial(\rho C_p T)}{\partial \tau}. \end{gathered} \tag{7.28}$$

For the equation to reflect interaction between internal and external parameters, it is necessary to separate internal and external parameters in it, i.e. internal and external processes and their interactions. Then the equation will be rewritten in the following form:

$$\rho_i C_{p_i} \operatorname{div}(V_i, T_i) - \lambda_i \nabla^2 T_i + \alpha_i \left(T^i_{wall} - T^e_{surr}\right) + \sum_i^n w^i_{P_i} \Delta H^i_{A_i} +$$
$$+ \rho_e C_{p_e} \operatorname{div}(V_e, T_e) - \lambda_e \nabla^2 T_e + \alpha_e \left(T^i_{wall} - T^e_{surr}\right) + \tag{7.29}$$
$$+ \sum_i^n w^e_{P_i} \Delta H^e_{A_i} = -\left[\frac{\partial(\rho_i C_{p_i} T_i)}{\partial \tau} + \frac{\partial(\rho_e C_{p_e} T_e)}{\partial \tau}\right].$$

Type of the process – spontaneous or non-spontaneous – may be determined with the help of the sign, which must be introduced before the corresponding term. Then the equation will be presented in the following view:

$$\pm \rho_i C_{p_i} \operatorname{div}(\upsilon_i, T_i) \mp \lambda_i \nabla^2 T_i \pm \alpha_i \left(T^i_{wall} - T^e_{surr}\right) \pm \sum_i^n w^i_i \Delta H^i_{A_i} \pm$$
$$\pm \rho_e C_{p_e} \operatorname{div}(\upsilon_e, T_e) \mp \lambda_e \nabla^2 T_e \pm \alpha_e \left(T^i_{wall} - T^e_{surr}\right) \pm \sum_i^n w^e_i \Delta H^e_{A_i} = \tag{7.30}$$
$$= \mp\left[\frac{\partial(\rho_i C_{p_i} T_i)}{\partial \tau} + \frac{\partial(\rho_e C_{p_e} T_e)}{\partial \tau}\right].$$

The sing *plus* determines spontaneous processes, and the sign *minus* – non-spontaneous ones. We also may transit from the heat flow equation to the entropy flow one.

7.4. Entropy Flow

State of the open flow exchanging energy in the heat form and substance with the surrounding is worthy consideration. In this flow (for example, flow of warm water in the Gulf Stream, air flows of tornado type, etc.) energy and substance exchange proceeds permanently along the whole length and volume of it. For the sake of convenience of transition from the flow of energy in the heat form, presented in the form of equation (7.30), to the flow of entropy, let us first present it as follows:

$$\pm \upsilon_i \left\{ \frac{\partial_i Q}{\partial x} + \frac{\partial_i Q}{\partial y} + \frac{\partial_i Q}{\partial z} \right\} \mp \left\{ \frac{\partial_i^2 Q}{\partial x^2} + \frac{\partial_i^2 Q}{\partial y^2} + \frac{\partial_i^2 Q}{\partial z^2} \right\} \pm \left(Q_{wall}^i - Q_{surr}^e \right) \pm$$
$$\pm \sum w_i Q_{A_i}^i \mp \upsilon_i \left\{ \frac{\partial_e Q}{\partial x} + \frac{\partial_e Q}{\partial y} + \frac{\partial_e Q}{\partial z} \right\} \mp \left\{ \frac{\partial_e^2 Q}{\partial x^2} + \frac{\partial_e^2 Q}{\partial y^2} + \frac{\partial_e^2 Q}{\partial z^2} \right\} \pm \quad (7.31)$$
$$\pm \left(Q_{surr}^e - Q_{wall}^i \right) \pm \sum w_e Q_{A_i}^e = \mp \left\{ \frac{\partial_i Q}{\partial \tau} + \frac{\partial_e Q}{\partial \tau} \right\}.$$

Let us divide this equation by T:

$$\pm \upsilon_i \left\{ \frac{\partial_i \frac{Q}{T}}{\partial x} + \frac{\partial_i \frac{Q}{T}}{\partial y} + \frac{\partial_i \frac{Q}{T}}{\partial z} \right\} \mp \left\{ \frac{\partial_i^2 \frac{Q}{T}}{\partial x^2} + \frac{\partial_i^2 \frac{Q}{T}}{\partial y^2} + \frac{\partial_i^2 \frac{Q}{T}}{\partial z^2} \right\} \pm \left(\frac{Q_{wall}^i}{T} - \frac{Q_{surr}^e}{T} \right) \pm$$
$$\pm \sum w_i \frac{Q_{A_i}^i}{T} \mp \upsilon_i \left\{ \frac{\partial_e \frac{Q}{T}}{\partial x} + \frac{\partial_e \frac{Q}{T}}{\partial y} + \frac{\partial_e \frac{Q}{T}}{\partial z} \right\} \mp \left\{ \frac{\partial_e^2 \frac{Q}{T}}{\partial x^2} + \frac{\partial_e^2 \frac{Q}{T}}{\partial y^2} + \frac{\partial_e^2 \frac{Q}{T}}{\partial z^2} \right\} \pm \quad (7.32)$$
$$\pm \left(\frac{Q_{surr}^e}{T} - \frac{Q_{wall}^i}{T} \right) \pm \sum w_e \frac{Q_{A_i}^e}{T} = \mp \left\{ \frac{\partial_i \frac{Q}{T}}{\partial \tau} + \frac{\partial_e \frac{Q}{T}}{\partial \tau} \right\}.$$

In this equation index i determines the internal heat exchange, and index e – the external one as before. These equations determine non-stationary exchange by reduced heat between internal and external systems, as well as the exchange inside each system with consideration for conditions of spontaneous and non-spontaneous proceeding of processes and their interaction with each other.

Analysis of this equation represents a quite complicated task, however, to ground it, it may be noted that change of the reduced heat proceeds in the same way as the change of energy in the heat form – via convective scattering of energy, heat transmission and mass transfer, exo- and endoprocesses.

These components of change of the reduced heat in interacting internal and external flows show themselves under stationary and non-stationary conditions.

From the reduced heat we may transit to entropy, and equation (7.32) can be rewritten as follows:

$$\pm \upsilon_i \left\{ \frac{\partial_i S_{con}}{\partial x} + \frac{\partial_i S_{con}}{\partial y} + \frac{\partial_i S_{con}}{\partial z} \right\} \mp \left\{ \frac{\partial_i^2 S_\lambda}{\partial x^2} + \frac{\partial_i^2 S_\lambda}{\partial y^2} + \frac{\partial_i^2 S_\lambda}{\partial z^2} \right\} \pm \left(S_{i,wall} - S_{e,surr} \right) \pm$$

$$\pm \sum w_i S_{i,chr} \mp \upsilon_e \left\{ \frac{\partial_e S_{con}}{\partial x} + \frac{\partial_e S_{con}}{\partial y} + \frac{\partial_e S_{con}}{\partial z} \right\} \mp \left\{ \frac{\partial_e^2 S_\lambda}{\partial x^2} + \frac{\partial_e^2 S_\lambda}{\partial y^2} + \frac{\partial_e^2 S_\lambda}{\partial z^2} \right\} \pm$$

$$\pm \left(S_{e,surr} - S_{i,wall} \right) \pm \sum w_e S_{e,chr} = \mp \left\{ \frac{\partial_i S}{\partial \tau} + \frac{\partial_e S}{\partial \tau} \right\}. \tag{7.33}$$

Here *con* and *chr* indexes mark convection and chemical reactions, respectively.

Hence, equation (7.33) displays different amount of entropy and its role in peculiar processes and the general one of interaction of internal and external systems.

In the entropy flow, the role of each component in total distribution of entropy in the flow and the surrounding changes permanently. In separate parts of the flow, the effect of scattering of energy and substance with prevailing growth of entropy over processes of its decrease under the effect of external factors prevails. In other parts, entropy decreases in the flow. This decrease of entropy prevails over its grow. Self regulation of entropy values *i* and *e* happens in the flow.

Chapter 8

The Helmholtz Energy for Spontaneous and Non-Spontaneous Processes

8.1. Analytical Expression of the Helmholtz Energy

Analytical expression of the Helmholtz energy is used for calculation of the free energy of spontaneous, equilibrium and non-spontaneous processes proceeding at constant temperature. In a particular case, it can be shown that volume of the system must be constant, too.

Analytical expression of the Helmholtz energy can be obtained on the basis of consolidation of equations of the first and the second laws of thermodynamics:

$$dQ = dU + dW \quad (8.1) \text{ and } TdS \overset{>}{\underset{<}{=}} dQ. \quad (8.2)$$

Signs *greater*, *equal* and *smaller* reflect spontaneous, equilibrium and non-spontaneous processes of entropy change in the system. Moreover, spontaneous process prevails over non-spontaneous one in the isolated system. All three cases may appear in open systems: prevailing of a spontaneous process over non-spontaneous one, quasi-static process (but not equilibrium one), prevailing of non-spontaneous process over spontaneous one.

Substituting (8.1) into equation (8.2), we obtain the main equation of thermodynamics in the general form:

$$TdS \overset{>}{\underset{<}{=}} dU + dW. \quad (8.3)$$

It should be noted that work of the equilibrium process differs from that, produced under conditions of thermodynamically irreversible conduction of the process. The work of equiponderantly proceeding process is maximal at $T = const$. If spontaneous process is conducted irreversibly, producing the work W_{sp}, less work is obtained than in the equilibrium process. In the case, if the work is spent for conduction of a non-spontaneous process (compression, heat delivery, etc.), work of the equilibrium process will be smaller than that of non-spontaneous, thermodynamically irreversible one $W_{n\text{-}sp}$. These two processes may be presented by the following inequalities:

$$dW_{equi} > dW_{sp}, \tag{8.4}$$

$$dW_{equi} < dW_{n\text{-}sp}, \tag{8.5}$$

where *sp* and *n-sp* are indexes of spontaneous and non-spontaneous processes, respectively. Hence, equation (8.3) and other ones, deduced on its basis, become much clear.

Let separate work from equation (8.3):

$$dW \underset{<}{\overset{>}{=}} TdS - dU. \tag{8.6}$$

Let integrate this expression in the range from the 1st to the 2nd state:

$$\int_1^2 dW \underset{>}{\overset{<}{=}} T\int_1^2 dS - \int_1^2 dU. \tag{8.7}$$

After integration, in accordance with Euler rules, we obtain the equation in finite differences of parameters:

$$W \underset{>}{\overset{<}{=}} T(S_2 - S_1) - (U_2 - U_1). \tag{8.8}$$

Grouping variables with the same indexes, we obtain:

$$W \underset{>}{\overset{<}{=}} (U_1 - TS_1) - (U_2 - TS_2). \tag{8.9}$$

The expression in brackets is marked by letter *A* and is called the Helmholtz energy:

$$A = U - TS. \tag{8.10}$$

Substituting (8.10) into (8.9), we obtain the expression, which connects energy change with the process work at $T = const$:

$$W \underset{>}{\overset{<}{=}} A_1 - A_2 . \tag{8.11}$$

From this expression the following values for relations between work and change of the Helmholtz energy can be determined:

$$\Delta A_{equi} > \Delta W_{sp}, \tag{8.12}$$

$$\Delta A_{equi} < \Delta W_{n\text{-}sp}, \tag{8.13}$$

$$A_1 - A_2 = -\Delta A_{equi} = W_{equi}. \tag{8.14}$$

These equations add inequalities (8.4) and (8.5).

Inequality (8.12) shows that at spontaneous proceeding of thermodynamically irreversible process a part of internal energy is lost as a consequence of various disturbances in the working substance (whirls, convectional flows, fluctuations of substance and energy, etc.). All this decreases useful free energy. Inequality (8.13) displays that the amount of work or energy, spent for changing of the system energy, exceeds the amount of work in the equilibrium process due to the same disturbances in the system. This decreases store of the free energy, which is accumulated by the system at consumption of work or energy.

8.2. Relation Between the Helmholtz Energy and Temperature and Volume

Let differentiate equation of the Helmholtz energy:

$$A = U - TS \tag{8.15}$$

and obtain the following expression:

$$dA = dU - TdS - SdT. \tag{8.16}$$

Let present work in the form of work of expansion:

$$dW = PdV. \tag{8.17}$$

Then the main equation of thermodynamics can be presented in a peculiar form:

$$TdS \underset{>}{\overset{<}{=}} dU + PdV . \tag{8.18}$$

Substituting (8.18) into (8.16), we get:

$$dA \underset{>}{\overset{<}{=}} -SdT - PdV. \tag{8.19}$$

In this equation, change of the Helmholtz energy is the function of temperature and volume.

Let substitute partial derivatives of the Helmholtz energy by T and V of the following view:

$$-S = \left(\frac{\partial A}{\partial T}\right)_V, \quad -P = \left(\frac{\partial A}{\partial V}\right)_T \tag{8.20}$$

and obtain the following expression:

$$dA \underset{>}{\overset{<}{=}} -\left(\frac{\partial A}{\partial T}\right)_V dT - \left(\frac{\partial A}{\partial V}\right)_T dV. \tag{8.21}$$

This equation enables change of the Helmholtz energy to be calculated by dependencies:

$$S = S(V;T),\ P = P(V;T)\ (n = 1 \text{ mole}). \tag{8.22}$$

8.3. Relation Between the Helmholtz Energy, Heat and Work

Differentiating the equation of the Helmholtz energy

$$A = U - ST, \tag{8.23}$$

we obtain full differential of the function:

$$dA \underset{>}{\overset{<}{=}} dU - TdS - SdT. \tag{8.24}$$

Substituting expression (8.3) into this equation, we get the following equation for the Helmholtz energy:

$$dA \underset{>}{\overset{<}{=}} -TdS - dW. \tag{8.25}$$

In this equation, the Helmholtz energy is bound to heat and work. From equation (8.25), two cases can be separated:

a) The following relation is obtained at $T = const$:

$$(\partial A)_T \underset{>}{\overset{<}{=}} -dW, \tag{8.26}$$

b) At $W = 0$ we obtain:

$$(\partial A)_W \underset{>}{\overset{<}{=}} -TdS. \tag{8.27}$$

These relations determine the change of the Helmholtz energy in various processes – spontaneous, non-spontaneous, equilibrium, non-equilibrium, ergoprocesses and entropy processes under various conditions of their conduction.

8.4. Criteria Determining the Type of the Process Proceeding

In a particular case, changing of the Helmholtz energy depends on temperature and volume in accordance with equation (8.19). Under constant T and V, in the absence of other kinds of work, the right part of this equation will be smaller, equal or greater than zero, we obtain the following relation:

$$(\partial A)_{T,V} \underset{>}{\overset{<}{=}} 0. \tag{8.28}$$

In fact, this relation determines that changing of the energy under constant T and V may change at the change of other parameters of the process only – pressure, composition, etc. Graphically, this relation can be presented by a curve with a minimum, as shown in Figure 8.1.

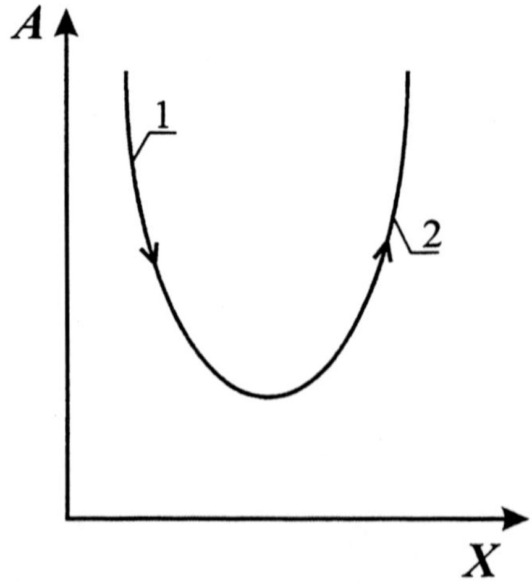

Figure 8.1. Changing of the Helmholtz energy with the change of an arbitrary parameter *X*, except temperature and volume.

Part 1 of the curve displays that value of the Helmholtz energy decreases, and this is characterized by the inequality reflecting spontaneous proceeding of the process of the following view:

$$(\partial A)_{T,V} < 0 . \tag{8.29}$$

The minimum on the curve displays that the first derivative from the Helmholtz energy equals to zero:

$$(\partial A)_{T,V} = 0 . \tag{8.30}$$

The right part 2 of the curve displays growth of the Helmholtz energy, which is possible only for non-spontaneous processes, if energy or work is expensed. This branch of the curve is characterized by the first derivative of the Helmholtz energy being greater than zero:

$$(\partial A)_{T,V} > 0 . \tag{8.31}$$

Consequently, the type of the process – spontaneous or non-spontaneous – may be determined by the sign of the derivative from the Helmholtz energy. In its turn, equality of the first derivative from the Helmholtz energy to zero points out the equilibrium existing between the system and the surrounding. The type of extremum is determined by the sign of the second derivative from the Helmholtz energy, which in the present case is positive:

$$(\partial^2 A)_{T,V} > 0 . \tag{8.32}$$

In addition, it should be noted that inequalities (8.29) and (8.31) determine the prevailing type of the process, because both spontaneous and non-spontaneous processes proceed simultaneously and permanently in any systems. Inequality (8.29) determines that a spontaneous prevails over a non-spontaneous one, in the present case, the latter playing the role of decelerating process. Inequality (8.31) determines prevailing of a non-spontaneous process over a spontaneous one, in this case, the latter being the decelerating one, which decreases the rate of proceeding of the non-spontaneous process, the rate of energy excess accumulation in the system comparing with its average level.

8.5. The Helmholtz Energy Including Heat of Irreversible – Spontaneous and Non-Spontaneous – Processes

For open systems, the reduced heat can be presented in the form of internal d_iQ and external heat exchange d_eQ. Analytically, this can be presented in the form as follows:

$$\frac{dQ}{T} = \mp\frac{d_iQ}{T} \pm \frac{d_eQ}{T}, \tag{8.33}$$

the sign *plus* determining heat delivery from the surrounding to the system, and the sign *minus* – heat elimination from the system to the surrounding. For internal heat exchange, the first term can be presented in the form of two reduced heats, which determine heat exchange between two parts of the system:

$$\frac{d_iQ}{T} = \sum\frac{d_iQ_i'}{T_i'} - \sum\frac{d_iQ_i''}{T_i''}, \tag{8.34}$$

where Q' and Q'' are amounts of heat, transmitted from the first substance to the second one and from the second substance to the system wall, which then transforms into value Q_e (heat transmission from the system wall into the surrounding).

Then entropy of the system interacting with the surrounding will be written down as follows:

$$dS = \pm\left(\sum d_iS' - \sum d_iS''\right) \mp d_eS. \tag{8.35}$$

If heat is transmitted from the surrounding to the system, its amount decreases in the surrounding (-), and grows in the system (+); this can be reflected by a sequence of signs (+) and (-). The first process is non-spontaneous; in the system, interacting with the surrounding, spontaneous process decelerates non-spontaneous one.

If the system eliminates heat to the surrounding, the amount of heat grows in it (+), and the system is cooled down (-). These processes can be determined by a sequence of signs (+) and (-). In this case, the prevailing process in the system will be a spontaneous one, decelerated by non-spontaneous process of the surrounding.

For these two cases, the Helmholtz energy can be presented by the following equations (at combination of equations of the first and the second laws of thermodynamics):

$$dA = -SdT - PdV - d_eQ + d_iQ, \tag{8.36}$$

$$dA = -SdT - PdV + d_eQ - d_iQ, \tag{8.37}$$

and at T and $V = const$ we obtain:

$$(\pm dA) \underset{>}{\overset{<}{=}} \mp d_iQ \pm d_eQ. \tag{8.38}$$

The inequality determines that for spontaneously proceeding processes the Helmholtz energy is consumed for transformation of it into heat of internal processes and for its transmission into the surrounding. For non-spontaneously proceeding processes, it is necessary for heat, delivered from outside, Q_e, to be higher than heat of internal processes, Q_i, i.e. $Q_e > Q_i$. Then the Helmholtz energy inside the system grows and, generally, will be higher comparing with the initial state of the system, interacting with the surrounding.

8.6. Connection of the Helmholtz Energy with Heat and Work

The Helmholtz energy for open systems may be expressed in the following form:

$$(\pm dA) \underset{>}{\overset{<}{}} \mp d_iQ \pm d_eQ \mp d_iW \pm d_eW, \tag{8.39}$$

where the sign *minus* determines decrease of work and energies, and the sign *plus* – growth of work and heat or exceeding of work and heat, or delivery of heat and work to the system. These inequalities reflect various types of combinations of the processes. In the case of prevailing of the processes, which cause decrease of the Helmholtz energy, processes with decrease and growth of heat and work may proceed in the system, but by value they exceed the flow of heat and work from the surrounding. In the second case, the flow of heat and work, delivered to the system from the surrounding, exceeds the flow from the system, which will lead to a decrease of the free energy store and to decrease of entropy in the surrounding.

In nature, processes of both types proceed simultaneously, and this determines proceeding of permanent changes in the surrounding and the system.

8.7. Analytical Expression of the Helmholtz Energy for Open Systems

8.7.1. Analytical Expression of the Helmholtz Energy for Open Systems with External and Internal Mass Exchange

Open systems represent systems, in which processes inside and between subsystems, inside the system, exchange of substance and energy between the system and subsystems and the surrounding proceed. These exchanges proceed thermodynamically irreversibly, with simultaneous participation of spontaneous and non-spontaneous processes in them. For more simplicity of consideration of processes, only two subsystems can be separated in the system, as shown in Figure 8.2.

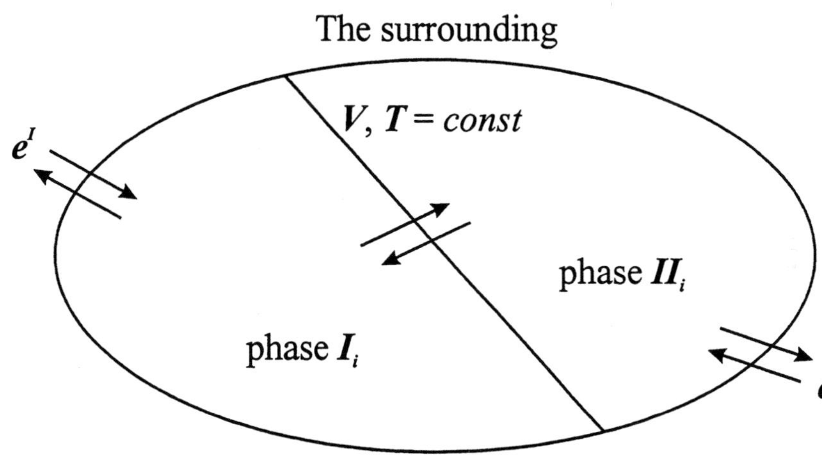

Figure 8.2. System including two subsystems, which exchange energy and substance with each other and with the surrounding.

The Helmholtz energy for an open system, interacting with the surrounding, can be presented in the form of the sum of energies for subsystems and the surrounding:

$$(\pm dA) \underset{>}{\overset{<}{=}} \mp d_i A^{\mathrm{I}} \mp d_i A^{\mathrm{II}} \pm d_e A, \tag{8.40}$$

where d_iA^{I}, d_iA^{II}, d_eA are Helmholtz energies for the 1st and the 2nd subsystems and the surrounding, respectively. Helmholtz energies include the ones, which are exchanged between subsystems and the surrounding, i.e. include change of entropy and the internal energy due to exchange of substance between subsystems and subsystems with the surrounding.

On the first stage of consideration of such systems, we neglect chemical interaction of molecules of substances with each other. For each type of exchanges the equation of the Helmholtz energy can be presented in the following view:

$$d_i A^{\mathrm{I}} = d_i \Phi^{\mathrm{I}} - T_i^{\mathrm{I}} d_i S^{\mathrm{I}} + \sum_j {}_i\mu_j^{\mathrm{I}} dn_j^{\mathrm{I}}, \tag{8.41}$$

$$d_i A^{\mathrm{II}} = d_i \Phi^{\mathrm{II}} - T_i^{\mathrm{II}} d_i S^{\mathrm{II}} + \sum_j {}_i\mu_j^{\mathrm{II}} dn_j^{\mathrm{II}}, \tag{8.42}$$

$$d_e A = d_e \Phi - T_e d_e S + \sum_j {}_e\mu_j^e dn_j^e, \tag{8.43}$$

where $d_i\Phi^{\mathrm{I}}$, $d_i\Phi^{\mathrm{II}}$, $d_e\Phi$ are energies, which subsystems (or phases) exchange with each other and with the surrounding, respectively; ${}_iS^{\mathrm{I}}$, ${}_iS^{\mathrm{II}}$, ${}_eS$ are entropy values, which subsystems exchange with each other and with the surrounding, respectively; ${}_i\mu^{\mathrm{I}}$, ${}_i\mu^{\mathrm{II}}$, ${}_e\mu$ are chemical potentials of substances in the 1st and the 2nd phases and in the surrounding, respectively; $n_j^{\mathrm{I}}, n_j^{\mathrm{II}}, n_j^e$ are amounts of substances in the 1st and the 2nd phases and in the surrounding, which may participate in exchanges.

In equation (8.43), functions of energy Φ and entropy S can be expanded into two parameters:

$$d_e\Phi = d_e\Phi^{\mathrm{I}} + d_e\Phi^{\mathrm{II}}, \tag{8.44}$$

$$d_eS = d_eS^{\mathrm{I}} + d_eS^{\mathrm{II}}, \tag{8.45}$$

Substituting (8.44) and (8.45) into (8.43), we obtain:

$$d_e A = d_e \Phi^{\mathrm{I}} + d_e \Phi^{\mathrm{II}} - T_e^{\mathrm{I}} d_e S^{\mathrm{I}} - T_e^{\mathrm{II}} d_e S^{\mathrm{II}} + \sum_j {}_e\mu_j^e d_e n_j^e. \tag{8.46}$$

Substituting (8.41), (8.42) and (8.46) into (8.40), we obtain the general expression for the Helmholtz energy dA for processes of exchange of energy, entropy and mass of substances between the system (subsystems) and the surrounding:

$$\begin{aligned} dA = {} & d_i \Phi^{\mathrm{I}} - T_i^{\mathrm{I}} d_i S^{\mathrm{I}} + \sum_j {}_i\mu_j^{\mathrm{I}} dn_j^{\mathrm{I}} + d_i \Phi^{\mathrm{II}} - T_i^{\mathrm{II}} d_i S^{\mathrm{II}} + \sum_j {}_i\mu_j^{\mathrm{II}} dn_j^{\mathrm{II}} + \\ & + d_e \Phi^{\mathrm{I}} + d_e \Phi^{\mathrm{II}} - T_e^{\mathrm{I}} d_e S^{\mathrm{I}} - T_e^{\mathrm{II}} d_e S^{\mathrm{II}} + \sum_j {}_e\mu_j^{\mathrm{I}} d_e n_j^{\mathrm{I}} + \sum_j {}_e\mu_j^{\mathrm{II}} d_e n_j^{\mathrm{II}} \end{aligned} \tag{8.47}$$

Grouping appropriate parameters for the system and the surrounding, we obtain the following expression:

$$\begin{aligned}(\pm dA)&=\pm d_i\Phi^{\mathrm{I}}\mp d_e\Phi^{\mathrm{I}}\pm d_i\Phi^{\mathrm{II}}\mp d_e\Phi^{\mathrm{II}}-\\&-\left(\pm T_i^{\mathrm{I}}d_iS^{\mathrm{I}}\mp T_e^{\mathrm{I}}d_eS^{\mathrm{I}}\pm T_i^{\mathrm{II}}d_iS^{\mathrm{II}}\mp T_e^{\mathrm{II}}d_eS^{\mathrm{II}}\right)+\\&+\left(\pm\sum_j {}_i\mu_j^{\mathrm{I}}d_in_j^{\mathrm{I}}\mp\sum_j {}_e\mu_j^{\mathrm{I}}d_in_j^{\mathrm{I}}\pm\sum_j {}_i\mu_j^{\mathrm{II}}d_in_j^{\mathrm{II}}\mp\sum_j {}_e\mu_j^{\mathrm{II}}d_in_j^{\mathrm{II}}\right).\end{aligned} \tag{8.48}$$

Admitting existence of the equality:

$$\begin{aligned}{}_en_j^{\mathrm{I}}+{}_in_j^{\mathrm{I}}&=n_j^{\mathrm{I}}\\{}_en_j^{\mathrm{II}}+{}_in_j^{\mathrm{II}}&=n_j^{\mathrm{II}},\end{aligned} \tag{8.49}$$

and obtaining derivatives

$$\begin{aligned}d_in_j^{\mathrm{I}}&=d_en_j^{\mathrm{I}}\\d_in_j^{\mathrm{II}}&=d_en_j^{\mathrm{II}},\end{aligned} \tag{8.50}$$

equation (8.48) will be rewritten as follows:

$$\begin{aligned}(\pm dA)&=\left(\pm d_i\Phi^{\mathrm{I}}\mp d_e\Phi^{\mathrm{I}}\right)+\left(\pm d_i\Phi^{\mathrm{II}}\mp d_e\Phi^{\mathrm{II}}\right)-\\&-\left(\pm T_i^{\mathrm{I}}d_iS^{\mathrm{I}}\mp T_e^{\mathrm{I}}d_eS^{\mathrm{I}}\right)-\left(\pm T_i^{\mathrm{II}}d_iS^{\mathrm{II}}\mp T_e^{\mathrm{II}}d_eS^{\mathrm{II}}\right)-\\&-\left(\pm\sum_j {}_i\mu_j^{\mathrm{I}}\mp\sum_j {}_e\mu_j^{\mathrm{I}}\right)d_in_j^{\mathrm{I}}-\left(\pm\sum_j {}_i\mu_j^{\mathrm{II}}\mp\sum_j {}_e\mu_j^{\mathrm{II}}\right)d_in_j^{\mathrm{II}}.\end{aligned} \tag{8.51}$$

In this equation, the upper sign determines non-spontaneous processes, and the lower sign – spontaneous ones.

8.7.2. Change of the Helmholtz Energy for Open Systems with Proceeding of Mass-Exchange Processes and Chemical Reactions in Them

In an open system and in the surrounding, chemical reactions may proceed, simple or complex reactions proceeding in each of subsystems (phases), and the phases exchange substances with each other and with the surrounding. To simplify equations, which describe chemical processes in the system and the surrounding, let us introduce a notion of chemical variable λ for reactions, proceeding in the phase 1, phase 2 and the surrounding *e*, and chemical affinity in the form of W^{I}, W^{II}, W_e. Then equation (8.51) may be presented in the following form:

$$
\begin{aligned}
(\pm dA) &= \left(\pm d_i\Phi^{I} \mp d_e\Phi^{I}\right) \mp \left(\pm T_i^{I} d_i S^{I} \mp T_e^{I} d_e S^{I}\right) - \\
&- \left(\pm \sum_j {}_i\mu_j^{I} \mp \sum_j {}_e\mu_j^{I} dn_j^{I} \mp W_i^{I} d\lambda_i^{I}\right) + \left(\pm d_i\Phi^{II} \mp d_e\Phi^{II}\right) - \\
&- \left(\pm T_i^{II} d_i S^{II} \mp T_e^{II} d_e S^{II}\right) - \left(\left[\pm \sum_j {}_i\mu_j^{II} \mp \sum_j {}_e\mu_j^{II}\right] dn_j^{II} \mp W_e^{II} d\lambda_e^{II}\right).
\end{aligned}
\tag{8.52}
$$

Differentiating equation (8.52) by time τ, we obtain the expression of rate for the change of the Helmholtz energy due to change of work (energy Φ), entropy, mass exchange and chemical reaction:

$$
\begin{aligned}
\left(\pm \frac{dA}{\partial\tau}\right) &= \left(\pm \upsilon_{\Phi_i}^{I} \mp \upsilon_{\Phi_e}^{I}\right) - \left(\pm T_i^{I} \upsilon_{S_i}^{I} \mp T_e^{I} \upsilon_{S_e}^{I}\right) - \left(\pm \sum_j {}_i\mu_j^{I} \mp \sum_j {}_e\mu_j^{I}\right) \upsilon_{n_j}^{I} \mp \\
&\mp W_i^{I} \frac{d\lambda_i^{I}}{\partial\tau} + \left(\pm \upsilon_{\Phi_i}^{II} \mp \upsilon_{\Phi_e}^{II}\right) \pm \left(\pm T_i^{II} \upsilon_{S_i}^{II} \mp T_e^{II} \upsilon_{S_e}^{II}\right) - \\
&- \left(\pm \sum_j {}_i\mu_j^{II} \mp \sum_j {}_e\mu_j^{II}\right) \upsilon_{n_j}^{II} \pm W_i^{II} \frac{d\lambda_i^{II}}{\partial\tau} \mp W_e^{I} \frac{d\lambda_e^{I}}{\partial\tau} \pm W_e^{II} \frac{d\lambda_e^{II}}{\partial\tau},
\end{aligned}
\tag{8.53}
$$

where υ is the rate of the appropriate process proceeding; λ is the chemical variable.

This equation (8.53) reflects the rate of approaching the process to the equilibrium in systems under the conditions of prevailing of spontaneous processes or the rate of removal of the process from the equilibrium under the conditions of prevailing of non-spontaneous processes. Subsidence of the Helmholtz energy with time in the system resulting energy and substance exchange inside the system between subsystems and between subsystems and the surrounding will be determined by negative value of the derivative. This displays that deceleration of the processes, proceeding spontaneously in the system, by non-spontaneous ones, proceeding in the system itself and in the surrounding, does not completely compensate the former one.

Positive sign of the derivative ∂A/∂τ points out that non-spontaneous processes, proceeding in the surrounding, by values of work and energy consumption exceeds those, given by spontaneous processes, conducted in the system.

Chapter 9

THE GIBBS ENERGY IN THERMODYNAMICALLY IRREVERSIBLE PROCESSES

The Gibbs energy is one of the most customary functions in thermodynamics, because it can be obtained in the form of simple dependencies under application of natural variables. Moreover, natural variables for this function P, T and others are easily measured with the help of appropriate devices, which simplifies quantitatives calculations of the Gibbs energy.

9.1. ANALYTICAL EXPRESSION FOR THE GIBBS ENERGY FOR CONDITIONS OF *P* AND *T* CONSTANCY IN CHEMICAL AND PHASE HOMOGENEOUS SYSTEM

The Gibbs energy is free under constancy of natural variables, which include P and T, the number of moles of substances, chemical variable, etc. It is calculated for processes, which proceed thermodynamically reversibly. The same function can be used for calculation of thermodynamically irreversible processes, subdivided into spontaneous and non-spontaneous ones.

The simplest expressions for the Gibbs energy are obtained for chemically homogeneous system, which includes a single phase only (the phase homogeneous system). The following equations were obtained on the basis of formulations of the first and the second law of thermodynamics:

$$dQ = dU + dW, \tag{9.1}$$

$$dQ \underset{>}{\overset{<}{=}} TdS, \tag{9.2}$$

$$dW = dW' + PdV, \tag{9.3}$$

where W and W' are the maximum and maximally useful work of the process conduction. The work of equilibrium proceeding process is maximal at $T = const$. At spontaneous irreversible conduction of the process producing the work W_{sp}, less amount of work is obtained, than in equilibrium conducted process. In the case, if the work is consumed for performance of non-spontaneously proceeding process (compression, heat delivery, etc.), the amount of work of the equilibrium process will be smaller than that of non-spontaneous, thermodynamically irreversible process $W_{n\text{-}sp}$. These two types of processes can be presented by the following inequalities:

$$dW_{equi} > dW_{sp}, \tag{9.4}$$

$$dW_{equi} < dW_{sp}, \tag{9.5}$$

where indexes *sp* and *n-sp* mark spontaneous and non-spontaneous processes, respectively. Hence, equations (9.1)-(9.3) and other ones, composed on its basis, become better understandable.

Let separate work from equation (9.3):

$$dW \underset{>}{\overset{<}{=}} TdS - dU . \tag{9.6}$$

Let integrate this expression in the range from the 1st to the 2nd state (chemical affinity). These equations become more complicated, if the following sorts of work are taken into calculation in them: surface tension forces - σds, forces of charge transfer in electric field - φdq; load lifting in the Earth gravitation field – $Mgdh$; work of osmotic pressure - πdV, where σ is the surface tension, s is the surface square, q is the charge, M is the body weight, g is the free dropping acceleration, h is the height, π is the osmotic pressure, V is the solvent volume. Grouping equations (9.1), (9.2) and (9.3), we obtain the following equation:

$$TdS \underset{>}{\overset{<}{=}} dU + PdV + dW' . \tag{9.7}$$

Then the maximal useful work must be separated from this equation:

$$dW' \underset{>}{\overset{<}{=}} TdS - dU - PdV . \tag{9.8}$$

Integrating this equation in the range from the 1st to the 2nd state, we obtain:

$$W' \underset{>}{\overset{<}{=}} T(S_2 - S_1) - (U_2 - U_1) - P(V_2 - V_1). \tag{9.9}$$

Then variables with the same indexes are grouped, after which the following difference of two combinations of parameters is obtained:

$$W' = (U_1 + PV_1 - TS_1) - (U_2 + PV_2 - TS_2). \tag{9.10}$$

The sum of parameters in brackets are marked by the letter G and are called the Gibbs energy:

$$G_1 = U_1 + PV_1 - TS_1 \text{ and } G_2 = U_2 + PV_2 - TS_2. \tag{9.11}$$

Substituting (9.11) into (9.10), we obtain the difference of Gibbs energies, equal to the maximal useful work for equilibrium processes, or relations in the form of inequalities for spontaneous and non-spontaneous processes:

$$W' \overset{<}{\underset{>}{=}} G_1 - G_2 \tag{9.12}$$

or

$$W' \overset{<}{\underset{>}{=}} -\left(G_2 - G_1\right), \tag{9.13}$$

$$W' \overset{<}{\underset{>}{=}} -\Delta G. \tag{9.14}$$

These expressions display that in equiponderantly proceeding processes the maximal useful work equals to diminution of the Gibbs energy or to increment of it. If the process proceeds thermodynamically irreversibly, the maximal useful work (the work of the equilibrium process) in spontaneous processes will be greater than the change of the Gibbs energy:

$$W'^{sp}_{irrev} < W'_{equi}, \tag{9.15}$$

because some part of the energy is scattered uselessly into the surrounding.

If the process proceeds thermodynamically irreversibly, with work delivery from outside, which transforms into the Gibbs energy change, the change of the Gibbs energy of the equilibrium process will be smaller than that in the irreversible one, because additional work is delivered to the system in order to overcome various perturbations in the system, i.e.

$$W'^{sp}_{irrev} > W'_{equi}. \tag{9.16}$$

9.2. Connection of the Gibbs Energy with Pressure and Temperature

Pressure and temperature are natural variables, at changing of which a change of the Gibbs energy is possible – reversibly and irreversibly, in spontaneous and non-spontaneous processes. In this case, it is necessary to take into account the following explanation. Spontaneous and non-spontaneous processes in any systems proceed simultaneously, interconnected and permanently. However, they cannot proceed purely, only prevailing of spontaneous over non-spontaneous processes or vice versa is possible. Pressure and temperature for the Gibbs energy enable connection to be set between them via derivatives in the simplest and clear view, and this, usually, is of sufficient practical usefulness.

First of all, the connection between the Gibbs energy, P, T and work can be set using the main equation of thermodynamics in the form:

$$TdS \overset{>}{\underset{<}{=}} dU + PdV + dW' \tag{9.17}$$

and the general expression of the Gibbs energy

$$G = U + PV - TS. \tag{9.18}$$

Differentiating equation (9.18), we obtain:

$$dG = dU + PdV + VdP - TdS - SdT. \tag{9.19}$$

Substitution of equation (9.17) into (9.19) gives the expression, connecting the Gibbs energy to the sought parameters:

$$dG \overset{<}{\underset{>}{=}} VdP - SdT - dW'. \tag{9.20}$$

At constant P and T (P, T = *const*) the expression, directly connecting any work, excluding the work of the working substance expansion, in the form of equality and in equalities is:

$$dG \overset{<}{\underset{>}{=}} - dW'. \tag{9.21}$$

This expression confirms expressions (9.12)-(9.16).

For systems, in which the work of potential forces only can be performed and thermal processes only can proceed, the equation for the Gibbs energy will obtain the following view:

$$dG \overset{<}{\underset{>}{=}} VdP - SdT. \tag{9.22}$$

Selecting equilibrium processes only, equation (9.22)can be rewritten in the following simplest view:

$$dG = VdP - SdT. \tag{9.23}$$

Quite simple expressions for first derivatives, connecting the Gibbs energy to natural variables can be deduced from this equation:

$$\left(\frac{\partial G}{\partial P}\right)_T = V, \tag{9.24}$$

$$\left(\frac{\partial G}{\partial T}\right)_P = -S. \tag{9.25}$$

Equation (9.24) determines that the first derivative from the Gibbs energy by pressure at constant temperature defines the volume. The measure of the Gibbs energy change with temperature growth under constant pressure is diminution of entropy. These derivatives allow notice that at selected independent variables, the Gibbs energy acquires properties of characteristic functions.

Second derivatives from the Gibbs energy by temperature and pressure determine changing of physical and thermal propertiess of the system and will possess the following structure:

$$\left(\frac{\partial^2 G}{\partial P \partial T}\right)_{T,P} = \left(\frac{\partial V}{\partial T}\right)_P = \frac{V}{V}\left(\frac{\partial V}{\partial T}\right)_P = V\alpha_T \overset{<}{\underset{>}{=}} 0. \tag{9.26}$$

$$\left(\frac{\partial^2 G}{\partial T \partial P}\right)_{P,T} = -\left(\frac{\partial S}{\partial P}\right)_T = -\frac{C_P}{T}\left(\frac{\partial T}{\partial P}\right)_T \overset{<}{\underset{>}{=}} 0. \tag{9.27}$$

$$\left(\frac{\partial^2 G}{\partial P^2}\right)_T = \left(\frac{\partial V}{\partial P}\right)_T = -\alpha_V V \overset{<}{\underset{>}{=}} 0. \tag{9.28}$$

$$\left(\frac{\partial^2 G}{\partial T^2}\right)_P = -\left(\frac{\partial S}{\partial T}\right)_P = -\frac{\partial(C_P T)}{T\partial T} = -\frac{C_P}{T} \underset{>}{\overset{<}{=}} 0. \tag{9.29}$$

9.3. Relation Between the Gibbs Energy, Heat and Work

In the general form, the Gibbs energy is determined by the equation of the following view for homogeneous systems:

$$G = U + PV - TS. \tag{9.30}$$

After differentiation of this expression, we obtain:

$$dG \underset{>}{\overset{<}{=}} dU + PdV + VdP - TdS - SdT. \tag{9.31}$$

Substituting the generalized equation into it, it is obtained:

$$TdS \underset{>}{\overset{<}{=}} dU + PdV + dW', \tag{9.32}$$

where W' is the maximal useful work for equilibrium processes and the work for irreversible processes, from which the work of expansion is separated. Substituting (9.32) into (9.31), we obtain simpler differential equation, containing natural variables and work:

$$dG \underset{>}{\overset{<}{=}} VdP - SdT - W'. \tag{9.33}$$

Under constant pressure, the entropy term in this equation can be substituted by heat:

$$dG \underset{>}{\overset{<}{=}} VdP - dQ_P - W'. \tag{9.34}$$

The following three consequences yield from this equation:

1. In the absence of heat exchange and conducted (consumed) work, changing of the Gibbs energy will be equal to the change of the potential energy:

$$dG \overset{<}{\underset{>}{=}} VdP\,. \tag{9.35}$$

2. Under constant pressure and in the absence of work, changing of the Gibbs energy will be bound to the heat exchange:

$$dG \overset{<}{\underset{>}{=}} -dQ_P\,. \tag{9.36}$$

In spontaneous processes under these conditions, the Gibbs energy may transform into heat, uselessly scattering into the surrounding. However, if work is delivered to the working substance, it may transform into a store of the free energy in the working substance under these conditions.

3. In the absence of heat exchange and at constant pressure, changing of the free energy will be bound to work:

$$dG \overset{<}{\underset{>}{=}} -W'\,. \tag{9.37}$$

In this case, changing of the free energy in spontaneous processes may be the source of work, and in non-spontaneous ones work may cause an increase of the Gibbs energy store in the working substance. These relations may become more complicated at proceeding of physical and chemical (or other) processes in the working substance and in the surrounding.

9.4. Changing of the Gibbs Energy in Isolated Systems

An isolated system does not exchange energy, work and substance with the surrounding. Spontaneous processes, which direct the system to the equilibrium, can proceed inside it only. Moreover, non-spontaneous processes in the closed system decelerate its approaching to the equilibrium. In the isolated system spontaneous processes prevail over non-spontaneous ones, and these processes proceed thermodynamically irreversibly. Under constant pressure and temperature value of the Gibbs energy in the system decreases, which is determined by inequality of the following view:

$$(\partial G)_{P,T} < 0\,. \tag{9.38}$$

Constancy of the Gibbs energy with time is the condition of the state of equilibrium reaching by the system, which is determined by the equality to zero of the first derivative:

$$(\partial G)_{P,T} = 0. \tag{9.39}$$

The minimal value of the Gibbs energy is characterized by the inequality that the second derivative from the Gibbs energy is greater than zero at P and $T = const$:

$$(\partial^2 G)_{P,T} > 0. \tag{9.40}$$

These inequalities are idealized, because it is impossible to obtain isolated (completely closed systems) in nature. They reflect the border behavior of the working substance in imagined systems.

9.5. Criteria of Types of Processes Proceeding in Open Systems

Open systems at constant P and T are able to exchange substance and work with the surrounding or change under the influence of other variables, which are not natural for the Gibbs energy. At constant P and T the following inequality can be deduced from equation (9.22):

$$(\partial G)_{P,T} \overset{<}{\underset{>}{=}} 0. \tag{9.41}$$

In the general case, the essence of this derivative, at changing of volume or other variables, can be understood from the curve, presented in Figure 9.1.

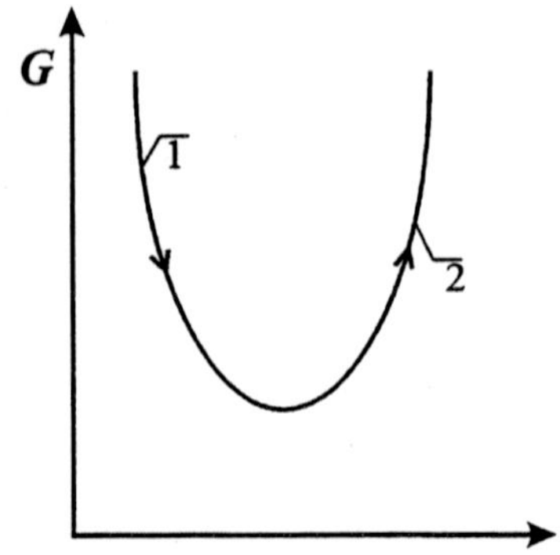

Figure 9.1. Changing of the Gibbs energy with arbitrary parameter X, not including P and T.

Slope of these curves determines the sign of the derivative. The left branch 1 of the curve possesses negative slope and defines decrease of the Gibbs energy with growth of parameter X. In the absence of external influence or at incomplete compensation of internal processes by external influences the Gibbs energy decreases, and this process proceeds spontaneously. Then for this case, the derivative will be negative also, i.e. it is smaller than zero:

$$(\partial G)_{P,T} < 0. \tag{9.42}$$

Leveling parameters of the system to these of the surrounding (for somewhat short period of time), the system and the surrounding reach the state of equilibrium, and the derivative for this state of the system with the surrounding will be equal to zero:

$$(\partial G)_{P,T} = 0. \tag{9.43}$$

This derivative determines the extremum of the Gibbs energy, the type of which is determined by the second derivative from the Gibbs energy, greater than zero:

$$\left(\partial^2 G\right)_{P,T} > 0. \tag{9.44}$$

At quite high compensation of internal spontaneous processes in the system by non-spontaneous ones proceeding in the surrounding, the Gibbs energy will grow in the system, i.e. the store of the free energy at P and $T = const$. Processes and their proceeding will be characterized by the following inequality:

$$(\partial G)_{P,T} > 0. \tag{9.45}$$

The rate of proceeding of spontaneous and non-spontaneous processes depends on the difference:

$$\left(\frac{\partial G}{\partial \tau}\right)_{P,T}^{sp} - \left(\frac{\partial G}{\partial \tau}\right)_{P,T}^{n-sp} \begin{matrix} < \\ = \\ > \end{matrix} \delta. \tag{9.46}$$

Each of these processes decelerates another and determines the total rate of proceeding of thermodynamic process, as well as the direction of the prevailing process, which proceeds in the system and in the surrounding.

9.6. The Gibbs Energy and its Connection with Heats of the System and the Surrounding

For natural variables, the Gibbs energy is bound to them by the equation of the following view:

$$dG \overset{<}{\underset{>}{=}} VdP - TdS\,. \tag{9.47}$$

This equation neglects the work. For open systems, entropy can be presented in the form of totality of entropies of internal parts (phases) of the system and entropy of the surrounding, between which permanent exchange proceeds:

$$dS = \sum d_i S' - \sum d_i S'' + d_e S\,. \tag{9.48}$$

Multiplying each term of entropy by temperature, we obtain the following expression:

$$TdS = \sum T_i' d_i S' - \sum T_i'' d_i S'' + T_e d_e S\,. \tag{9.49}$$

Work of potential forces can be also separated into potential works of internal parts of the system and of the surrounding:

$$VdP = V_i dP_i + V_e dP_e\,. \tag{9.50}$$

Let introduce notions of heats of internal and external processes into equation (9.49):

$$TdS = \Sigma d_i Q' - \Sigma d_i Q'' + d_e Q. \tag{9.51}$$

Substituting (9.51) and (9.50) into (9.47), we obtain:

$$dG \overset{<}{\underset{>}{=}} V_i dP_i + V_e dP_e - \left(\sum d_i Q' - \sum d_i Q''\right) - d_e Q\,. \tag{9.52}$$

9.7. Connection of the Gibbs Energy to Heat and Work

Heat exchange and work production processes can proceed inside open systems. Moreover, the system is able to exchange heat and work with the surrounding. Then changing of the Gibbs energy may be presented by the following expression:

$$(\pm dG) \underset{>}{\overset{<}{=}} \pm \sum V_i dP \mp \sum V_e dP \pm \sum d_i Q \mp \sum d_e Q \pm \sum d_i W \mp \sum d_e W. \quad (9.53)$$

This equation neglects mass exchange between phases inside the system and the surrounding, as well as chemical transformations of reagents inside and between the phases and in the surrounding.

9.8. Changing of the Gibbs Energy for Open Systems with Mass Exchange and Chemical Reactions. Analytical Expression of the Gibbs Energy with External and Internal Mass Exchange

In an open system under constant pressure and temperature, both heat exchange and mass exchange may proceed inside the system and between the system and the surrounding. Then changing of the Gibbs energy will include changing of it between phases in the system and in the surrounding:

$$dG = d_i G^{\mathrm{I}} + d_i G^{\mathrm{II}} + d_e G, \quad (9.54)$$

where $d_i G^{\mathrm{I}}$, $d_i G^{\mathrm{II}}$, and $d_e G$ are changes of the Gibbs energy for phases in the system and in the surrounding, respectively.

Analytical expression of the Gibbs energy can be presented in the form of difference of enthalpy and TS term:

$$G = H - TS. \quad (9.55)$$

Differentiation of this expression gives:

$$dG \underset{>}{\overset{<}{=}} dH - TdS - SdT. \quad (9.56)$$

At $T = const$ we obtain:

$$dG \underset{>}{\overset{<}{=}} dH - TdS. \quad (9.57)$$

In accordance with formula (9.54), equation (9.57) for phases inside the system and the surrounding can be presented as follows:

$$\begin{aligned}(\pm dG)\overset{<}{\underset{>}{=}}\pm\sum d_iH'\pm\sum d_iH''\mp\sum T_i'd_iS_i'\mp\sum T_i''d_iS_i''\mp\\ \mp\sum d_eH^e\mp\sum T_edS^e.\end{aligned} \tag{9.58}$$

In this equation specific features of interaction of each phase with the surrounding are not separated, but spontaneous and non-spontaneous processes are.

9.9. Analytical Expression of the Gibbs Energy for Open Systems with Mass Exchange Between Phases and the Surrounding

In open systems, mass exchange processes may proceed between phases inside the system and the surrounding through permeable interlayer or closing layers, as, for example, in flows of air masses, water thermal flows. For the sake of convenience, expression of enthalpy for the system can be presented in the form of two terms – energy and work:

$$d_iH^{\mathrm{I}} = d_i\Phi^{\mathrm{I}} + V^{\mathrm{I}}dP + \mu_i^{\mathrm{I}}dn_i^{\mathrm{I}}, \tag{9.59}$$

where Φ^{I} is the energy, by which the first phase exchange with the second one or phase with higher energy exchange with these of lower energy; μ_i^{I} is the chemical potential of the phases; n_i^{I} is the number of moles of the substance in these phases. Then in the general view, expression (9.59) will be rewritten in more complicated form:

$$d_iH^{\mathrm{I}} = \sum d_i\Phi^{\mathrm{I}} + \sum V^{\mathrm{I}}dP^{\mathrm{I}} + \sum \mu_i^{\mathrm{I}}dn_i^{\mathrm{I}}. \tag{9.60}$$

For phases with lower store of energy, the equation will be analogous, but with index II:

$$d_iH^{\mathrm{II}} = \sum d_i\Phi^{\mathrm{II}} + \sum V_i^{\mathrm{II}}dP^{\mathrm{II}} + \sum \mu_i^{\mathrm{II}}dn_i^{\mathrm{II}}. \tag{9.61}$$

Energy (enthalpy) for the surrounding can be easily presented in the usual form:

$$d_eH = \sum d_e\Phi^e + \sum V_e^i d_eP^e + \sum_j \mu_j^e d_en_j. \tag{9.62}$$

Total change of the Gibbs energy will take into calculation the process of mass and energy exchange between phases inside the system and with the surrounding, and the equation can be presented in the first approximation in the following view:

$$(\pm dG) \underset{>}{\overset{<}{=}} \pm d_i\Phi^{\mathrm{I}} \pm d_i\Phi^{\mathrm{II}} \pm d_e\Phi \pm \left(\sum V_i^{\mathrm{I}} dp^{\mathrm{I}} + \sum V_i^{\mathrm{II}} dp^{\mathrm{II}} + \sum V_i^e dP^e\right) \pm$$
$$\pm \frac{1}{2}\left(\sum \mu_i^{\mathrm{I}} d_i n_i^{\mathrm{I}} + \sum \mu_i^{\mathrm{II}} d_i n_i^{\mathrm{II}}\right) \pm \left(\frac{1}{2}\sum \mu_i^{\mathrm{I}} d_i n_i^{\mathrm{I}} + x\sum \mu_e d_e n_i\right) \pm \qquad (9.63)$$
$$\pm \left(\frac{1}{2}\sum \mu_i^{\mathrm{II}} d_i n_i^{\mathrm{II}} + (1-x)\sum \mu_e d_e n_i\right).$$

Basing on this equation, it is possible to select conditions, under which influence of energy and work on the system from the surrounding prevails, that determines the prevailing of non-spontaneous processes proceeding over spontaneous ones, and vice versa. In living organisms, basing on self-development processes during all their life, non-spontaneous processes prevail over spontaneous ones, whereas in dead organisms spontaneous processes prevail already over non-spontaneous ones. At the same time, delivery of energy and substance proceeds in both cases, but it is of different quality and ability to be consumed. Hence, in living organisms spontaneous and non-spontaneous processes proceed simultaneously, however, their ratio changes abruptly at the moment of its origination (prevalence of non-spontaneous processes over spontaneous ones at the sacrifice of feeding from the outside), as well as at the moment of death (spontaneous processes prevail over non-spontaneous ones). The same reasoning can be also applied to technological systems.

9.10. The Gibbs Energy for Open Systems with Mass Exchange and Chemical Reactions

For open systems, in which chemical reactions proceed, and under the condition that chemical reactions proceed in the surrounding using substances from the system, the equation for calculation of the Gibbs energy can be deduced, broadening equation (9.63). Adding the product of chemical affinity and chemical variable λ to equation (9.63), we obtain the expression for calculation of the Gibbs energy:

$$
\begin{aligned}
(\pm dG) \overset{<}{\underset{>}{=}} & \pm d_i\Phi^{\mathrm{I}} \pm d_i\Phi^{\mathrm{II}} \pm d_e\Phi \pm \left(\sum V_i^{\mathrm{I}} dp^{\mathrm{I}} + \sum V_i^{\mathrm{II}} dp^{\mathrm{II}} + \sum V_i^{e} dp^{e}\right) \pm \\
& \pm \frac{1}{2}\left(\sum \mu_i^{\mathrm{I}} d_i n_i^{\mathrm{I}} + \sum \mu_i^{\mathrm{II}} d_i n_i^{\mathrm{II}}\right) \pm \left(\frac{1}{2}\sum \mu_i^{\mathrm{I}} d_i n_i^{\mathrm{I}} + x\sum \mu_e d_e n_i\right) \pm \\
& \pm \left(\frac{1}{2}\sum \mu_i^{\mathrm{II}} d_i n_i^{\mathrm{II}} + (1-x)\sum \mu_e d_e n_i\right) \pm \frac{1}{2}\left(\sum \nu_i^{\mathrm{I}} \mu_i^{\mathrm{I}} + \sum \nu_i^{\mathrm{II}} \mu_i^{\mathrm{II}}\right) d\lambda \pm \\
& \pm \left(\frac{1}{2}\sum \nu_i^{\mathrm{I}} \mu_i^{\mathrm{I}} + \sum \nu_e \mu_e\right) d\lambda \pm \left(\frac{1}{2}\sum \nu_i^{\mathrm{II}} \mu_i^{\mathrm{II}} + (1-x)\sum \nu_e \mu_e\right) d\lambda.
\end{aligned}
\tag{9.64}
$$

This equation takes into calculation energy exchange in both the form of heat and in another form between phases inside the system and with the surrounding, as well as mass exchange and chemical interaction of substances in the phases and the surrounding.

This expression can be differentiated by time, which will give expression for rates of the open system approach to the state of equilibrium or removal the state of the system and the surrounding from it:

$$
\begin{aligned}
\left(\pm \frac{dG}{d\tau}\right) = & \pm\left(\Phi_i^{\mathrm{I}}\right)' \pm \left(\Phi_i^{\mathrm{II}}\right)' \mp (\Phi_e)' \pm \sum V_i^{\mathrm{I}} w_{P_i}^{\mathrm{I}} \pm \sum V_i^{\mathrm{II}} w_{P_i}^{\mathrm{II}} \mp \sum V_e w_{P_e} \pm \\
& \pm \frac{1}{2}\left(\sum \mu_i^{\mathrm{I}} w_{n_i}^{\mathrm{I}} + \sum \mu_i^{\mathrm{II}} w_{n_i}^{\mathrm{II}}\right) \pm \left(\frac{1}{2}\sum \mu_i^{\mathrm{I}} w_{n_i}^{\mathrm{I}} + x\sum \mu_e w_{n_e}\right) \pm \\
& \pm \left(\frac{1}{2}\sum \mu_i^{\mathrm{II}} w_{n_i}^{\mathrm{II}} + (1-x)\sum \mu_e w_{n_e}\right) \pm \frac{1}{2} W^{\mathrm{I-II}} w_\lambda \pm W^{\mathrm{I}-e} w_\lambda \mp \\
& \mp W^{\mathrm{II}-e} w_\lambda,
\end{aligned}
\tag{9.65}
$$

where W is the chemical affinity, w_λ is the process rate.

9.11. Conclusion to the Theoretical Part

In the present view, this book determines a new direction in thermodynamics of irreversible processes. For thermodynamically irreversible processes, all processes can be conditionally subdivided into spontaneous and non-spontaneous ones. The former ones proceed under conditions, when the system possesses an excess of energy or higher order (excess of pressure, substance, and other factors), and at deficiency of compensation in open systems processes proceed uncontrollably – from less probable to higher probable state – minimum of the energy store, maximum of disorder, smaller size of the surface, etc.

If intensive factors in the surrounding exceed the same ones in the system, non-spontaneous processes will proceed. These processes (delivery of heat, work, energy, pressure boost, etc.) will cause growth of store of the free energy, substance, order, etc. in the system. In the first case, non-spontaneous process will decelerate spontaneous ones, and in the second case, spontaneous processes will decelerate or decrease efficiency of non-spontaneous ones. Ability to separate these pairs of processes at the interaction of the open system with the surrounding will allow deeper clearing out the essence of the processes proceeding, the mechanism of their interaction with each other, degree of their mutual influence. On this ground, more full mathematical descriptions of the processes could be made, method of interaction degree control of these two mutually excluding processes could be developed, and efficiency of application of the processes could be increased.

Chapter 10

PRACTICAL EXAMPLES OF INFLUENCE OF RELATION OF SPONTANEOUS AND NON-SPONTANEOUS PROCESSES ON TECHNOLOGICAL AND NATURAL PHENOMENA

10.1. DISPERSION OF THE CONTINUUM

Dispersion of solids and liquids concludes in decay of the lattice of solids and in overcoming forces of intermolecular interactions between atoms, ions or molecules in the continuum. Solids and liquids preserve their form due to cohesion forces, displayed between particles, as well as different specific interactions of atoms, ions and molecules with each other – covalent and ionic bonds, complexes with charge transfer, and specific interactions – Van-der-Waals forces. To decompose continuity of a solid or liquid, it is necessary to conduct an appropriate work against cohesion forces. This work determines non-spontaneous processes. When work W_{disp} affects a continuum, it is spent for decomposition of crystalline or amorphous lattice of a solid or for dispersion of a liquid. This work is spent for overcoming cohesion forces (chemical, crystallochemical, and other bonds, also including Van-der-Waals forces). In this case, particles of the dispersion phase appeared are characterized by external and internal surfaces, which determine action of adhesion forces between particles. Overcoming of cohesion forces due to effects of mechanical or any other work on the substance characterizes a non-spontaneous process. Adhesion forces determine ability of particles to adhere to each other and, in this case, decrease aggregate stability of dispersion systems. These forces determine proceeding of a spontaneous process of aggregation of particles in the dispersion system, and the processes are differently displayed in creation of the dispersion system. Consequently, mechanical (or any other) work delivered from outside is determined by difference of cohesion and adhesion forces:

$$W_{disp} = f_{coh} + f_{adh}, \qquad (10.1)$$

where f_{coh}, f_{adh} are cohesion and adhesion forces. In thermodynamically irreversible processes, this work equals to the change of enthalpy and entropy, according to the main equation of thermodynamics:

$$\Delta G_d = \Delta H - T\Delta S, \tag{10.2}$$

where $\Delta G_d = -W_{disp}$ is the change of the Gibbs energy, equal to the maximal useful work, spent for dispersion of the continuum.

The work for overcoming cohesion forces is spent for diminution of the volume of decomposing substance and increase of the surface of particles, which can be determined by the equation:

$$W_{coh} = f_{coh} = k_1 d^3 + k_2 \sigma d^2, \tag{10.3}$$

where k_1 and k_2 are proportionality coefficients, σ is the surface tension, d is the particle diameter. In this equation, the first term determines the work, spent for deformation of the substance in the volume, and the second one – the change of the particle surface square. When size of particles of the dispersion phase decreases, the external surface square of particles and store of excessive free energy, that characterizes adhesion of particles to each other.

Work of adhesion at interaction of particles with surface atoms of the same nature will be determined by the expression:

$$f_{adh} = -U_{AA} F n_s \gamma_s, \tag{10.4}$$

where U_{AA}, F, γ_s, n_s are the bond energy between homogeneous atoms, surface of particles, coordination number of atoms in the surface layer and the number of bonds of two interacting surfaces, respectively.

This energy determines spontaneously proceeding processes of aggregation of small particles into large ones due to existence of large excessive energy of particles of the dispersion phase with high dispersion degree. Substituting (10.3) and (10.4) into (10.1), we obtain the equation, uniting spontaneous and non-spontaneous processes:

$$W_{disp} = k_1 d^3 + k_2 \sigma d^2 - U_{AA} F n_s \gamma_s. \tag{10.5}$$

Analyzing this equation, it could be noted that at low dispersion degree of the phase in the dispersion system first two terms are greater, than the third one, and the main role is played by the non-spontaneous process in dispersion of the continuum. The third term determines low deceleration of crushing of the solid or the liquid. At quite highly developed surface, the third term prevails over the first and the second ones. The process of agglutination of particles already prevails and prevents crushing of the substance, and in this case, the process of crushing of particles stops. Aggregation of particles (spontaneous process) prevails in the system. Full aggregation of all particles in the system is prevented by a non-spontaneous process, which decelerates aggregation.

It could be noted that at compensation of a non-spontaneous process by a spontaneous one, changes in such dispersion system not proceed yet.

10.2. Thermal Machines

Thermal machines can be the source of work production or, vice versa, they can transmit heat from cold substances to heated up ones at the work consumption. The first machine is called direct one, and the second - reverse one. They can work between the same heater and cooler, as shown in Figure 4.3.

The following pairs of spontaneous and non-spontaneous processes can be separated in this pair of conjugated thermal machines. In the direct machine pairs can be separated in the way, shown in Table 10.1.

Table 10.1 Types of spontaneous and non-spontaneous processes.

Machine A		Machine B	
Process		Process	
Spontaneous	**Non-spontaneous**	**Spontaneous**	**Non-spontaneous**
1. Heat transmission from the heater to the working substance	1. Cooling down of the heater from T_{surr} to $T < T_{surr}$	1. Consumption of work by machine B	1. Heat transfer from the cooler to the working substance $T_{cooler} < T_{work.subst.}$
2. Expansion of gas in the machine	2. Compression of gas	2. Absorption of heat by the heater $T_{work.subst.} > T_{heater}$	2. Heat transmission from the working substance to the heater $T_{work.subst.} < T_{heater}$
	3. Transformation of heat into work		
	4. Transmission of heat to the cooler from the working substance		

The number of such stages can be finitely increased, if consider all processes in more detail, which proceed in the conjugated pair of thermal machines. The same example is only to illustrate the fact that both machines work due to permanent and inseparable interaction of spontaneous and non-spontaneous processes.

10.3. Mechanical Processes

In mechanical (force) interpretation, the notion of spontaneous processes may be spread upon throttle and external adiabatic processes. They proceed without effect of external parameters and with decrease of pressure. At the same time, external adiabatic processes, proceeding with pressure boost, may be related to non-spontaneous ones. Beside friction, this process is also decelerating one – spontaneous escape of gas from throttle slides.

10.4. Catalytic Processes

Catalysis is a complex physicochemical process, including a great number of conjugated spontaneous and non-spontaneous stages. Conditionally, the following conjugated processes can be selected:

Spontaneous processes	**Non-spontaneous processes**
1. Diffusion to the side of lower concentrations	1. Substance concentration increase
2. Deorientation of molecules on the catalyst surface	2. Favorable orientation of molecules in the field of active centers
3. Sorption of molecules on active centers	3. Desorption of molecules from active centers
4. Transition of exited molecules and active centers into the main state	4. Activation of molecules and active centers with transfer of electrons from UOMO to LVMO
5. Redistribution of electrons between atomic and molecular orbitals	5. Creation of electron pairs
6. Energy transmission from exited MO to VMO (vacant MO)	6. Desorption of molecules from active centers

Spontaneous processes proceed with decrease of the Gibbs energy and growth of entropy, and non-spontaneous ones – with growth of the Gibbs energy and decrease of entropy (in appropriate systems). The former processes affect the latter, causing appropriate changes of energy and entropy, and vice versa.

Combination of the Earth and the outer space may be considered as the system and the surrounding. The surrounding is characterized by relict temperature of 2 – 4 K. In relation to these conditions, general processes proceeding on the Earth may be considered as spontaneous and non-spontaneous ones from the point of view of their prevalence over one another.

Prevalence of spontaneous processes over non-spontaneous ones is typical for autumn-winter periods and vice versa, prevalence of non-spontaneous processes over spontaneous ones is typical for spring-summer periods.

APPENDIX

Chapter 11

EQUATIONS OF EQUILIBRIUM THERMODYNAMICS AND THE METHOD OF DETERMINATION OF THE PROCESS TYPE BASING ON THERMODYNAMICS OF SPONTANEOUS AND NON-SPONTANEOUS PROCESSES

11.1. EQUATIONS OF THE FIRST LAW OF THERMODYNAMICS AND THEIR CONNECTION WITH THE PROCESS TYPE

The equation of the first law of thermodynamics had been presented before in the form, considering sign, which reflect type of the process proceeding:

$$(\pm dQ) = \pm dU \pm PdV, \tag{11.1}$$

where $dW = PdV$ is the work connected with change of the volume of the working substance. In classic thermodynamics, the equation of the first law of thermodynamics is presented in less particular form:

$$dQ = dU + PdV. \tag{11.2}$$

This equation does not consider a possibility of interaction between the working substance and the surrounding and vice versa, and spontaneous and non-spontaneous processes could not be separated on its basis. It is possible to reduce from equation (11.2) to equation (11.1) basing on the method of Euler integrating of equation (11.2) in the range from the first to the second and from the second to the first state.

For the sake of convenience of analyzing equations, obtained below, let us introduce the following designations:

$+Q$ – heat is delivered to the working substance from the surrounding (non-spontaneous process),

$-Q$ – heat is eliminated from the working substance into the surrounding (spontaneous process),

$+\Delta U$ – rise of the internal energy of the working substance (non-spontaneous process),

$-\Delta U$ – diminution of the internal energy of the working substance (spontaneous process),

$+W$ – the working substance conducts work (spontaneous process),

$-W$ – work is consumed for changing properties of the working substance (non-spontaneous process, for example, adiabatic and isothermal gas compression).

11.2. The Equation of the First Law of Thermodynamics Determining Heat Delivery to the Working Substance

Heat may be delivered to the working substance from a source with higher store of energy at temperature higher, than that of the working substance and the surrounding. Such process is non-spontaneous, and heat delivered will be consumed for rising the store of the internal energy of the working substance (non-spontaneous process) and conduction of work (spontaneous process):

$$dQ_P^{surround} = dU + pdv\,, \tag{11.3}$$

where *surround* determines heat delivery from the surrounding.

Equation (11.3) allows a note that spontaneous and non-spontaneous processes proceed in the system interacting with the surrounding simultaneously, permanently and interconnectedly. Spontaneous process is decelerated by a non-spontaneous one and vice versa.

Non-spontaneous processes in equation (11.3) are heat delivery and growth of the internal energy, and spontaneous one is conduction of work by the system. Equation (11.3) determines prevalence of the non-spontaneous process over the spontaneous one. To separate these conditions, equation (11.3) must be integrated in the range from the first to the second state:

$$\int_1^2 dQ_P = \int_1^2 dU + \int_1^2 p dv. \tag{11.4}$$

In this equation, U and v are functions of the state. Consequently, heat is the function of the state, too. Finishing integration at $p = const$, we obtain integral equation of the first law of thermodynamics:

$$Q_{P_2} - Q_{P_1} = (U_2 - U_1) + p(v_2 - v_1). \tag{11.5}$$

From equation (11.5), it follows that $Q_{P_2} > Q_{P_1}$ and $U_2 > U_1$, i.e. energy delivery in the form of heat to the working substance and growth of the internal energy at heating up the working substance are non-spontaneous processes. They could not pass themselves, external influences on the working substance are necessary. At the same time, the working substance would have to increase volume spontaneously, due to which it produces work of changing the state in the surrounding.

In the form of finite differences, equation (11.5) can be presented in the following form:

$$\Delta Q_P = \Delta U + p\Delta v. \tag{11.6}$$

11.3. Equation of the First Law of Thermodynamics for Spontaneous Process

Spontaneous processes may proceed in the system in the case, when intensive parameters of the system (P, T, μ, etc.) are higher, than these of the surrounding. To select type of the equation of the first law of thermodynamics for spontaneous processes, prevailing over non-spontaneous ones, equation (11.3) should be integrated in the range from the first to the second state:

$$\int_2^1 dQ_P = \int_2^1 dU + \int_2^1 p dv. \tag{11.7}$$

Finishing integration at $p = const$, we obtain the following equation:

$$Q_{P_1} - Q_{P_2} = (U_1 - U_2) + p(v_1 - v_2). \tag{11.8}$$

In the form of finite differences, equation (11.8) will be rewritten in the following view:

$$-\Delta Q = -\Delta U - p\Delta v \tag{11.9}$$

or

$$-p\Delta v = -\Delta Q + \Delta U. \tag{11.10}$$

Equation (11.10) determines that work, spent for changing the state of the working substance (non-spontaneous process), is consumed for energy elimination into the surrounding (spontaneous process) and enlarging changes of the internal energy (non-spontaneous process).

In the general view, equation of the first law of thermodynamics in the frames of thermodynamics of spontaneous and non-spontaneous processes can be presented by finite differences with regard to appropriate signs:

$$(\pm dQ) = \pm dU \pm pdv\,. \tag{11.11}$$

As a consequence, the above-displayed equations are obtained from this one.

11.4. Changing of the Helmholtz Energy for Spontaneous and Non-Spontaneous Processes

11.4.1. Spontaneous Process

The main equation of thermodynamics in the general view reflects connection between the maximal work, entropy and internal energy:

$$dW = TdS - dU. \tag{11.12}$$

To display prevalence of spontaneous processes over non-spontaneous ones, let integrate equation (11.12) in the range from the first to the second state:

$$\int_W dW = \int_1^2 TdS - \int_1^2 dU\,. \tag{11.13}$$

Finishing integration, we obtain the following equation:

$$W = T(S_2 - S_1) - (U_2 - U_1). \tag{11.14}$$

In this equation, we group variables with the same indexes:

$$W = (U_1 - TS_1) - (U_2 - TS_2). \qquad (11.15)$$

Difference of two variables in brackets at $T = const$ is the Helmholtz energy A:

$$A_1 = U_1 - TS_1, \qquad (11.16)$$

$$A_2 = U_2 - TS_2. \qquad (11.17)$$

Substituting (11.16) and (11.17) into (11.15), we obtain difference of Helmholtz energies, equal to the maximal work:

$$W = A_1 - A_2. \qquad (11.18)$$

In this difference, the Helmholtz energy of the initial state of the system is higher, than that of the system state, i.e.

$$A_1 > A_2. \qquad (11.19)$$

Consequently:

$$A_1 - A_2 > 0, \text{ and } A_2 - A_1 < 0 \text{ and } \Delta A < 0. \qquad (11.20)$$

In the closed system, this inequality determines spontaneous proceeding of the process only, which proceeds up to reaching the state of equilibrium in the system. When such system interacts with the surrounding, spontaneous leveling of the Helmholtz energy of the system with that of the surrounding happens. Under this interaction, the system may become the source of production of the maximal work in thermodynamically reversible process. Then:

$$W^{reverse} = -(A_2 - A_1). \qquad (11.21)$$

Work of spontaneous irreversible process $W^{irreverse}$ will be smaller than the work of spontaneous one $W^{reverse}$, i.e.:

$$W^{irreverse} < W^{reverse}. \qquad (11.22)$$

This equation displays that the free energy of the working substance decreases and entropy increases with diminution of the Helmholtz energy.

Concluding this section, it can be noted that integration of equation (11.12) in the range from the first to the second state points out that under irreversible conditions the process proceeds spontaneously, and at quasi-equilibrium proceeding of the process, it also proceeds from the side of prevailing spontaneous process. In this case, spontaneous process is decelerated by non-spontaneous ones.

11.4.2. The Main Equation of Thermodynamics for Non-Spontaneous Processes

For non-spontaneous processes, which may proceed due to internal influence on the working substance only by work or excessive energy, the main equation of thermodynamics is integrated in the range from the second to the first state:

$$\int_{W} dW = T\int_{2}^{1} dS - \int_{2}^{1} dU . \tag{11.23}$$

After integration we obtain:

$$W = (U_2 - TS_2) - (U_1 - TS_1). \tag{11.24}$$

Substituting expressions for the Helmholtz energy into equation (11.24), we obtain that the maximal work of equilibrium physicochemical non-spontaneous process will correspond to growth of the difference of Helmholtz energies of the end A_2 and initial A_1 states of the working substance:

$$W = A_2 - A_1. \tag{11.25}$$

For non-spontaneous process, the difference of Helmholtz energies exceeds zero:

$$A_2 - A_1 > 0 \text{ and } A_2 > A_1 \tag{11.26}$$

and, consequently, the Helmholtz energy of the working substance at the end of the process (non-spontaneous) conducted is higher than that of the initial state of the working substance. For irreversible non-spontaneous process, the work consumed for changing the state of the system is greater than that of the reversible process:

$$W_{consum}^{irreverse} > W^{reverse} . \tag{11.25}$$

In conclusion, it may be noted that work (or energy) of non-spontaneous process, consumed by the surrounding, is spent for increase of the internal energy of the system and for decrease of entropy change.

In the general case, the sign of thermodynamic function change points out prevalence of spontaneous or non-spontaneous process. If calculation displays that the change of the Helmholtz energy is negative, spontaneous process prevails over non-spontaneous one in the system interacting with the surrounding. If the change of the Helmholtz energy is positive, non-spontaneous process prevails over spontaneous one. Moreover, in the first case, non-spontaneous process decelerates proceeding of spontaneous one, and vice versa in the second one.

Temporal changes of the Helmholtz energy may be presented in the form of algebraic sum of partial derivatives of A by time:

$$\frac{\partial \Delta A_{sp}}{\partial \tau} + \frac{\partial \Delta A_{n-sp}}{\partial \tau} = \frac{\partial \Delta A_{tot}}{\partial \tau}. \tag{11.28}$$

Using ideas of thermodynamics of spontaneous and non-spontaneous processes, more full notion about types of interaction of phase-open systems with the surrounding can be obtained. The above-discussed method may be applied to analysis of behavior of systems and the surrounding at other natural variables.

REFERENCES

1. Eremin E.N., *Grounds of Chemical Thermodynamics*, 1978, Moscow, Vysshaya Skola, 391 p. (Rus)
2. S. de Groot and Mazur P., *Non-equilibrium Thermodynamics*, 1964, Moscow, Mir, 456 p. (Rus)
3. Kolesnikov I.M., *Thermodynamics of Physicochemical Processes*, 1994, Moscow, GANG, 288 p. (Rus)
4. Kolesnikov I.M., Vinokurov V.A., Kolesnikov S.I., *Thermodynamics of Spontaneous and Non-spontaneous Processes as the Ground of Proceeding of Physicochemical Processes, Science and Technology of Hydrocarbons*, 1999, No. 2, pp. 78 – 83. (Rus)
5. Kolesnikov I.M., Kolesnikov S.I., *Main Laws of Thermodynamics and Stages of Its Development, Science and Technology of Hydrocarbons*, 1999, No. 6, pp. 40 – 44. (Rus)
6. Nickols G. and Prigozhin I.R., *Self-organization in Non-equilibrium Systems*, 1979, Moscow, Mir, 512 p. (Rus)
7. Prigozhin I.R., *Introduction into Thermodynamics of Irreversible Processes*, 1960, Moscow, Inostrannaya Literatura, 127 p. (Rus)

8. Reznitsky L.A. and Guzei A.S., *Energy of Ions Preference and Energetic of Coordination Polyhedrons, In: Chemical Thermodynamics and Equilibrium*, 1978, Moscow, VINITI, pp. 70 – 119. (Rus)
9. Frolov Yu.G., *The Course of Colloidal Chemistry, Surface Phenomena and Dispersion Systems*, 1988, Moscow, Khimia, 464 p. (Rus)
10. Cherkinsky Yu.S., *Global Thermodynamics*, 1994, Moscow, AO POLYEX, 504 p. (Rus)
11. Shchukin E.D., Pertsev A.V., and Amelina E.A., *Colloidal Chemistry*, 1982, Moscow, MSU, 348 p. (Rus)

INDEX

A

absolute temperature, 67, 70, 71, 74, 75, 85, 91
absolute vacuum, 3, 6, 7, 10, 13, 85, 96
affinity, 1, 128, 130, 141, 142

B

Benar cells, 21, 22

C

Carnot cycle, 66, 67, 68, 69, 70, 71, 75, 77
chemical affinity, 128
chemical potentials, 17, 36, 37, 102, 105, 126
chemical variable, 5, 36, 127, 128, 129, 141
closed systems, xii, xiii, 2, 45, 52, 136

D

density, 7, 14, 21, 54, 107, 108
diffusion coefficient, 14
Dispersion, viii, 145, 156

E

EC, 58, 59, 63, 64, 65, 66, 68, 69, 70, 71
efficiency coefficient, 58, 59, 63, 64, 65, 66, 70, 75
enthalpy, 7, 10, 17, 28, 34, 35, 36, 38, 43, 44, 47, 49, 139, 140, 146
equilibrium processes, 1, 3, 4, 20, 31, 51, 56, 57, 61, 78, 79, 92, 96, 131, 133, 134
Extensive parameters, 8

F

first law of thermodynamics, 32, 33, 78
flow of energy, 106, 113
flow of entropy, 106
flow of heat, 124
flow of substance, 14
Fourier law, 111
function of the state, 151

G

G.I. Gess law, 38
Gibbs energy, 7, 10, 129, 131, 132, 133, 134, 135, 136, 137, 138, 139, 141, 146, 148

H

heat capacity, 7, 14, 33, 34, 44, 46, 47, 101, 107, 108
heat conductivity, 14, 21, 107, 108, 109, 110
heat exchange, 12, 45, 47, 92, 111, 112, 114, 123, 134, 135, 139
heat flow, 94, 107, 108, 110, 111, 112, 113
heat transmission, 49, 51, 61, 94, 95, 106, 107, 111, 114, 123
Helmholtz energy, 7, 10, 18, 39, 117, 118, 119, 120, 121, 122, 124, 125, 126, 128, 153, 154, 155

I

interaction of spontaneous and non-spontaneous processes, 2, 24, 29, 52, 83, 148
internal energy, 5, 7, 8, 9, 10, 15, 17, 24, 28, 29, 30, 31, 32, 33, 34, 38, 43, 49, 50, 79, 80, 96, 103, 119, 126, 150, 151, 152, 155

[illegible] 41, 46, 47, 107, 109, 110, 111, 112, 129, [illegible]
[illegible] 108, [illegible]
[illegible] of thermal [illegible] series, 9, 15, 23, 27, 28, 31, 32, 33, 35, [illegible] 56, 57, [illegible] 66, 75, 78, 80, 83, 96, 97, 125, 149, 151, 152

M

magnetic field, 8, [illegible]
[illegible] 137, 139, 140, 142

[illegible]

[illegible] 36, [illegible]

second row of thermal [illegible]
self-organization, [illegible]
spontaneous and anti-spontaneous [illegible] 2, 3, 4, 8, 9, 10, 12, [illegible] 24, 25, 26, 28, 31, 33, 34, [illegible] 60, 62, 63, 65, [illegible] 130, 131, 133, [illegible] 147, 150, 153, 154